Encyclopedia of Plants and Flowers

ENCYCLOPEDIA OF PLANTS AND FLOWERS

Edited by:

N.C. Gautam

Delve
PUBLISHING

www.delvepublishing.com

Encyclopedia of Plants and Flowers

N.C. Gautam

Delve Publishing
224 Shoreacres Road
Burlington, ON L7L 2H2
Canada
www.delvepublishing.com
Email: orders@arclereducation.com

ISBN: 978-1-77469-389-6 (Hardcover)

ABOUT THE EDITOR

Prof. N.C. Gautam is M.Sc. Ph.D. in Botany-Genetics & Plant Breeding. He is a former Vice-Chancellor, M.G.C.G.V., State University Chitrakoot, Satna (M.P.) and VBS Purvanchal University, Jaunpur, UP. He has also worked as a Dean, Post graduate studies. NDUAT, Faizabad and Dean & Faculty of Hort., NDUAT, Faizabad. He has over 49 years experience in academics and more than 46 years of experience in teaching of Post Graduate and Under Graduate courses at University level. He also has 48 years experience, guided more than 8 Ph.D. students and 15 M.Sc Ag theses. He has published 20 books and many research papers. He is a Fellow of Indian society of Vegetable science, Fellow of Indian Botanical Society and Fellow of CHAI. He has received 15 awards and honors at national and international level. He has also worked on improving 30 Varieties of rice & vegetable crops besides other viable technology. He has also served as Chairperson/member of Expert Committee-UGC, Chairman/Member in Vice Chancellor Search Committees for many universities and Chairman-UGC and ICAR State Universities.

TABLE OF CONTENTS

LIST OF FIGURES

Figure 190: *Cistus albidus* is a xerophyte that grows in European countries such as France, and Italy and North African countries like Morocco.

Figure 191: *Satyrium carneum*. Ground orchid with typical zygomorphic floral anatomy

Figure 192. Zygote formation process

ABSTRACT

This Plants and Flowers book makes reference to 100s of plants and adorns its pages with great quality pictures, definitions and overall cultivation guidance. It's a truly delightful inventory in digital print. As an amateur studying florae, it is a priceless resource for understanding plant names. It's certainly not going to be comprehensive in a globe of 420,000 types of plant, also mentioning the massive numbers of cultivars and varieties.

Notwithstanding, it is loaded with the most common and contemporary domestic plants. The book is divided into different plant types, including trees, bushes, climbers, bulbs, water plants, perennials, annuals and others such as delicate plants and desert flora. It provides a close up of blossoms or leaves and gives a true sense of what the plant resembles. This book is an unquestionable resource for any genuine, novice or master gardeners. Flowering plants are a kind of vascular plant that produces blossoms to reproduce. These plants produce seeds inside a fruit product. The scientific term for blossoming plants is angiosperms.

PREFACE

Blossoming plants emerged on Earth around 120 million years ago, the Cretaceous era. That is much later in the geologic timeframe: If all of the globe's history were compacted into 60 minutes, blossoming plants would exist for just 90 seconds. In any case, when they took hold around 90 million years ago, they quickly differentiated into a variety of blossoming plant groups of the contemporary world. Today, blossoming plant species dwarf by twenty to one those of conifers and ferns, which had flourished for 200 million years before the original bloom appeared. As a food source blossoming plants give people and animals the sustenance that is key for survival. Truth be told, if it were not for blossoming plants, we wouldn't even exist. From palms and oaks to water lilies and wildflowers, across the stretch of cornfields and citrus plantations, blossoming plants have come to dominate the universe of plant science and agriculture.

Botanists call blossoming plants angiosperms, from the Greek terms "vessel" and "seed." Different from conifers, which create seeds in exposed cones, angiosperms encase their seeds in the fruit product. Each fruit product contains at least one carpel, hollow compartments that nourish the seeds. Cut a tomato down the middle, for example, and you'll see carpels. These products are the defining qualities of all angiosperms, which numbers around 235,000 species.

Exactly when and how did the original blossoming plants arise? Charles Darwin contemplated that question, and paleobotanists are as yet looking for a response. All through the 1990s discoveries of fossilized blossoms in Australia, Europe, Asia, and North America provided significant clues. Simultaneously, modern paleonbotany has undergone an explosion, not much the same as the Cretaceous flower eruption.

Presently, old-style fossil trackers with digging tools and magnifying instruments compare materials utilizing genetic sequencing to track contemporary plant families in reverse to their beginning points.

With a precise image of the development of a blossoming plant, it's critical to understand more about the construction and function of different plants which will assist with responding to specific queries: What kinds of species can it be crossed with? What kinds of pollinators are viable?" This, goes toward reasonable and useful techniques for cultivation, just as a more clear comprehension of the bigger course of development. Flowers and plants are significant in light of the fact that they help to clean the air. They produce oxygen and consume carbon-dioxide.

This book is evocative and lavishly illustrates a variety of blossoms and plants, which have given us such a lot of sustenance, and delight. The radiant yellow of marigold and the blissful red of geranium, the suggestive aroma of a lotus or a saffron-implanted paella—there is no limit to reasons to adore blossoms. Throughout centuries and across the globe, the book examines the abundance of botanicals throughout the ages; and in practically every religious, and social setting.

From the daffodils of spring to the last chrysanthemums of pre-winter, this is likewise a book about seasons. In detail and picture, it is a rich source of botanicals. Sunflower, rose, lily and poppy, —among others—have inspired meaning, value, and motivation from the start of time, and that is why they are indispensable to different societies.

A

Abscission

Abscission is either the natural or artificial separation of a plant from one of its organs. In most cases, abscission is done to plant organs that are no longer necessary in the plant's life. Leaf abscission is the natural shedding of leaf plants, most occurs based on a specific season. The main purpose for leaf abscission is to ensure that the plants retain water and nutrients within a plant structure, more so during the summer season. In situations where leaf abscission occurs in an unusual season other than it is expected, it is believed to be a sign of some type of plant disease (Zehnder et al., 2007).

Fruit abscission is the natural shedding of fruits from the plant. In most cases, fruit abscission is a sign of plant diseases, hence it is recommended that, the plants be analyzed to reduce further transmission of the disease to other plants, and also to plants around the infected one. However, fruit abscission could be due to excessive ripening of a plant part.

The elimination of unwanted parts of a plant is contributed by different factors. One, abscission can be artificial in order to eliminate infected parts of a plant, to limit transmission of the infectious disease to other parts of the plant. Flower abscission is the natural shedding of plant flowers, which occurs after pollination, since the flowers are no longer needed for the plant.

Generally, plant abscission is so important for it helps in improving plant growth; for instance, artificial abscission that involves the cutting of extra plant branches ensures that the plant has enough space around it, which promotes its rate of growth (Yamada et al., 2003). Eliminating the extra tree branches creates enough space for maximum accumulation of plant produce. It also promotes yield production, for instance, the cutting off of infected plant parts improves their rate of production.

Plant abscission, more natural abscission, needs to be controlled to maintain the plant at a health condition; since, for instance, excessive leaf abscission limits the rate of transpiration, which is such a crucial process in plants. However, in some situations, the rate of plant abscission might be low, requiring acceleration of the abscission process.

For ages, abscission has been being done in a manual way, by use of a sharp cutting equipment or anything close to that. However, the future of plant abscission promises to make abscission less time-consuming, and quite easy. It is being rumored that, in near future, artificial abscission, will rely on the use of chemicals.

Figure 1: A leaf abscission illustration.

Source:https://www.google.com/imgres?imgurl=https%3A%2F%2F4.bp.blogspot.com%2F-4PbmnaU98sk%2FWfOZyQ-piEI%2FAAAAAAAAAkjI%2Fpss2j5tiomkAdVM62Md0q6WfRGprrpUvACLcBGAs%2Fw1200-h630-p-k-no-nu%2FLeaf-abscission.jpg&imgrefurl=http%3A%2F%2Flifeofplant.blogspot.com%2F2011%2F03%2Fleaf-abscission.html&tbnid=y2HCgwdlq32k-M&vet=12ahUKEwjRkfeO--30AhUJXxoKHUrpDLsQMygDegUIARC0AQ..i&docid=bUr5IMWbb_E_AM&w=620&h=326&itg=1&q=leaf%20abscission%20in%20plants&ved=2ahUKEwjRkfeO--30AhUJXxoKHUrpDLsQMygDegUIARC0AQ

Anther

An anther, is part of a plant flower which is categorized under the plant's male reproductive system. It is sack-like in nature, a structure that helps it in holding the male gametes, referred to as pollen grains. The anther is supported a few inches high from the bottom of the flower by a structure, which is long, slender and stalk-like, referred to as the filament. Both the filament and the anther complete the male reproductive system of a plant. The holdings of pollen grains in an anther are referred to as pollen sacs (Yakhin et al., 2017). Pollen grains from pollen sacs are released during fertilization; whereby they are moved from the anther to the stigma, either on the same flower or a different flower; through a process referred to as pollination. The movement of pollen grains from an anther to the stigma of the same plant is referred to as self-pollination; whereas, the movement of pollen grains from an anther to a stigma of another plant is referred to as cross-pollination.

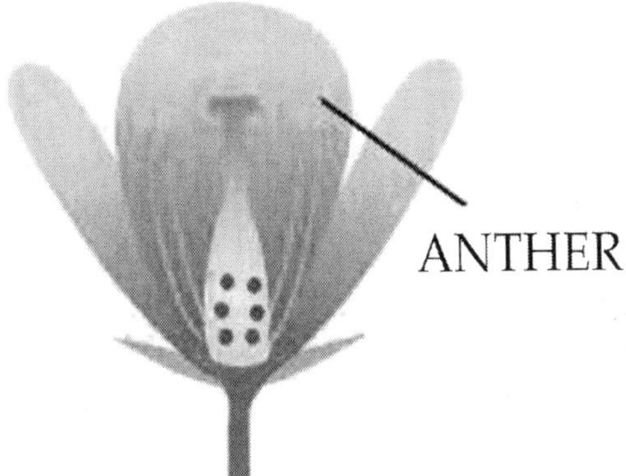

Figure 2: An image showing the location of an anther in a plant flower.

Source;https://www.google.com/imgres?imgurl=https%3A%2F%2Fimages.ctfassets.net%2Fvrrt8fsfwf0e%2FKwfGU3Cg9J0e5do4Gw3lD%2Fe338d3752c772e925904f6b13b7e5864%2FScreenshot__688_.png&imgrefurl=https%3A%2F%2Fwww.chegg.com%2Flearn%2Fbiology%2Fintroduction-to-biology%2Fanther&tbnid=AtVG3gwHYIsphM&vet=12ahUKEwiW0NO4mO_0AhVI8IUKHQbFBRUQMygEegUIARC9AQ..i&docid=6FrygJEhZ-btHM&w=247&h=167&itg=1&q=anthers%20in%20plants&ved=2ahUKEwiW0NO4mO_0AhVI8IUKHQbFBRUQMygEegUIARC9AQ

As stated earlier, an anther has a sack-like general structure; however, in details, the anther is knob-like, bi-lobed, a region where the connective tissue connects the two lobes into one. The lobes are further divided into two chambers each, totaling four chambers for a complete anther. An anther has an outer layer, referred to as an epidermis, which protects the internal parts of the anther. The internal structure of an anther is made up of three main layers. The outermost layer is referred to as, outer endothecium (Arruda et al., 2015). Outer endothecium is made up of flattened cells which further help in the protection of internal layers of the anther. The flattened cells undergo frequent differentiation, to help in strengthening the outer endothecium. The second layer, rather, the middle layer is composed of three more layers, constituted from parenchymatous cells. The innermost layer, referred to as the inner tapetum, is made up of pyramid-shaped cells. It acts as a nourishment layer, providing nourishment to the other two layers of the anther.

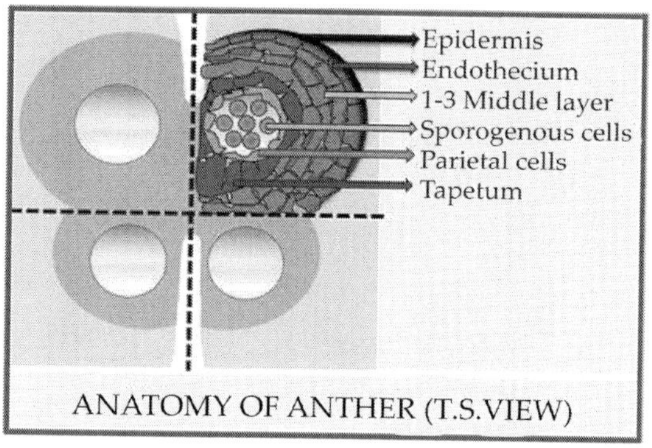

Figure 3: An image showing the internal structure of an anther.

Source:https://www.google.com/imgres?imgurl=https%3A%2F%2Fbiologyreader.com%2Fwp-content%2Fuploads%2F2019%2F08%2FAnatomy-of-anther.jpg&imgrefurl=https%3A%2F%2Fbiologyreader.com%2Fanther.html&tbnid=MrMMwi9WMycjAM&vet=12ahUKEwi0-OLpmO_0AhVtgM4BHWYu-AcUQMygAegUIARCqAQ..i&docid=7CHUJhPJkOe6KM&w=420&h=319&itg=1&q=internal%20structure%20of%20an%20anther&ved=2ahUKEwi0-OLpmO_0AhVtgM4BHWYuAcUQMygAegUIARCqAQ

Anthers are divided into two main categories; the first structure being based on their structure; and the second structure being based on their attachment to the filament. The first category, based on their structure, is further divided into 8 subdivisions. Bi-lobed anthers are the most common, and they include anthers made up of two lobes, which are clearly visible in mature anthers. Unilocular anthers, have a single lobe; however, in most cases, Unilocular anthers were once bi-lobed, only that, they ended up having a single lobe due to; the destruction of one lobe, or destruction of the connective tissue known to combine the two lobes. Linear anthers are common in the Acalypha class of plants. Rounded anthers appear in the Merculiaris class of plants (Yakhin et al., 2017). Sagittate anthers appear in the vinca class of plants. Sinuous anthers appear in the Cucurbitaceae plant species. Reniform anthers, as the name suggests, are reniform in nature. The last subdivision of anthers based on their structure is the appendiculate anthers, which are common in the Erica cinerea plant species.

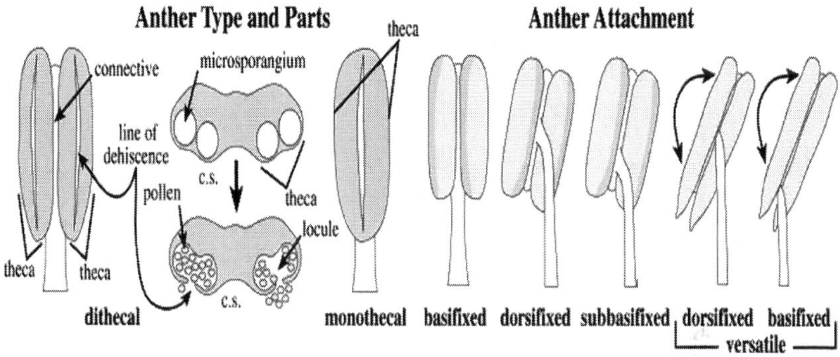

Figure 4: An image showing the classification of anthers based on their structure.

Source: https://www.sciencedirect.com/topics/immunology-and-microbiology/anther

The second category of anthers, based on their attachment to the filament, is further divided into four subdivisions. The adnate anthers, are located in between the lower part and upper part of a flower filament, more of a continuation of the filament. They are most common in waters lilies and Magnolia plants. Innate anthers, also referred to as the basifixed anthers, are attached to the filament at their basal region (Wisler & Norris, 2005). The dorsifixed anthers, are attached to the filament from their back side; and they are common in Sesbania plants. The last subdivision, is the versatile anther type, which is located exactly at the middle of a filament. The anther also attaches to the filament, from its middle position, allowing it to be able to swing freely.

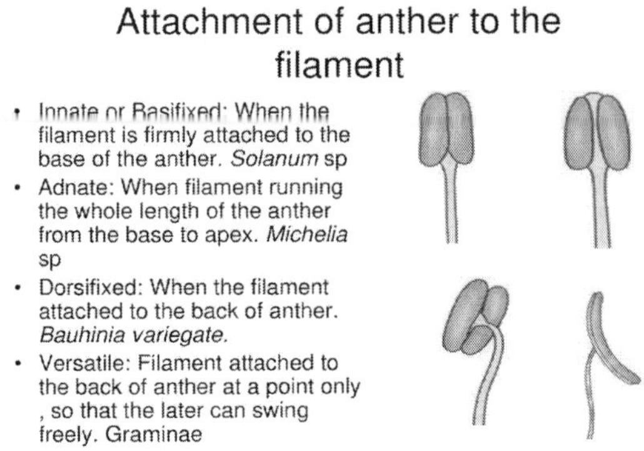

Figure 5: An image showing the classification of anthers based on their attachment to the filament.

Source: https://slideplayer.com/slide/17170501/

The importance of anthers to plant life depends on the different components which make up a complete anther. As stated earlier, the epidermis, serves as a protection to the anther, through the provision of structural integrity and strength. The epidermis also helps in air circulation, from the anther to the surrounding, and vice versa; preventing extreme conditions such as drying up of the anthers due to lack of sufficient refreshment. The endothecium, which as stated earlier, is the outermost internal layer of an anther, provides support to the other internal layers, and also offers additional protection to them, making it to be more of a complement to the epidermis (Weitemier et al., 2014). The middle layer of the epidermis, is used as a storage space for starch. The starch is later used to strengthen pollen grains, and make them healthy for purposes of fertilization in flowers. The innermost layer, the tapetum, provides nourishment to the entire anther, but more specifically, to the pollen grains, promoting their growth rate. Based on the flower type, the tapetum helps in the definition of the pollen grains compatibility, a factor that is crucial during fertilization.

Acaulescent

Acaulescent is a term used to refer to all plant species which do not have a stem, and if they do, the stem is not visible since it is very short. Acaulescent plants have the habit of growing their leaves close to the ground, covering the tiny stem present in them. Some of these plants have unique stems in the sense that, some part of the stem is underground, will all a small portion of it being visible above the ground; and they include the likes of tubers, bulbs, and rhizomes.

The list of acaulescent plants is quite large, since, the following are all categorized under acaulescent plants: Aloe, Agave, Yucca, Onions, Garlic, Chives, Crocus, Dandelions, Carrots, Pineapple, Cilantro, Lettuce, Spinach, some Cycad species, some Primrose species, some Oxalis plant species, and some Palm species, among others.

Figure 6: An image of an acaulescent plant.

Source:https://www.google.com/imgres?imgurl=http%3A%2F%2Fwww.the-dailygarden.us%2Fuploads%2F4%2F5%2F4%2F9%2F45493619%2Fpublished%2Fimg-3343.jpg%3F1561695244&imgrefurl=https%3A%2F%2Fwww.thedailygarden.us%2Fgarden-word-of-the-day%2Facaulescent&tbnid=6Gzehs7HxGk9RM&vet=12ahUKEwiaud_t0e_0AhUFpBoKHXMNCg8QMygAegUIARCuAQ..i&docid=7Ae5KT1UvRfUbM&w=735&h=551&itg=1&q=acaulescent%20plants&ved=2ahUKEwiaud_t0e_0AhUFpBoKHXMNCg8QMygAegUIARCuAQ

Accrescent

In most living organisms, growth is known to reach a limit depending on age. However, in some plant species, growth seems to be a never-ending pattern; or, in some cases, several parts of the plant might have reached their growth limit, leaving behind one plant part which continues to grow with an increase in plant age. Although, the continuous growth of a specific plant organ does reach its limit after a specific development stage has been reached.

There are different levels of accrescence, depending on whether it is a plant organ that is accrescent or the entire plant. The first level is that of a flower being accrescent. The plant flower is said to continue growing until it gets to its maturity level; however, it is not the flower that is continuously getting larger, rather its stem, which attaches it to the plant, or the flower leaves, the calyx (Wilkinson et al., 2012). This level is categorized under slightly accrescent plants. The second level of accrescent is that of plants which continue getting bigger with age. A good example of this accrescent level is exhibited by the Giant Sequoias plants.

The importance of understanding accrescence in plants is that it helps in the identification of unknown plant species; and in flowering plants, it helps in determining the right harvest time, based on the flower growth rate.

Figure 7: An image showing an accrescent part of a plant.

Source:https://www.google.com/imgres?imgurl=http%3A%2F%2Fwww.thedailygarden.us%2Fuploads%2F4%2F5%2F4%2F9%2F45493619%2Ftomatillo-1511524-1920_orig.jpg&imgrefurl=https%3A%2F%2Fwww.thedailygarden.us%2Fgarden-word-of-the-day%2Faccresence&tbnid=KAuj-klQwiMBgM&vet=12ahUKEwjz48Sg1u_0AhVNwoUKHR99A8UQMygAegQIARB9..i&docid=NQPSD2ISQErEVM&w=1067&h=800&itg=1&q=accrescent%20in%20plants&ved=2ahUKEwjz48Sg1u_0AhVNwoUKHR99A8UQMygAegQIARB9

Accumbent

Accumbent is a term used in plants to refer to plant parts lying against other plant parts; or, an entire plant lying against another plant or a certain surface. For instance, cotyledons could be said to be accumbent for they lie against the radicle on one of its edges.

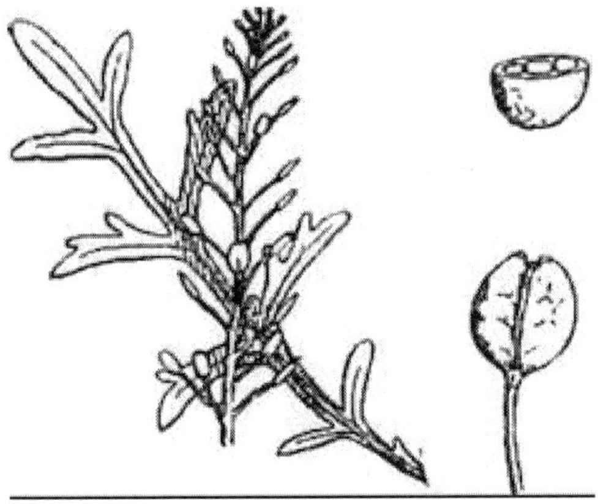

Figure 8: An illustration of accumbency in plants.

Source:https://www.google.com/imgres?imgurl=https%3A%2F%2Fc8.alamy.com%2Fcomp%2FPFYPT8%2Fgrays-new-manual-of-botany-a-handbook-of-the-flowering-plants-and-ferns-of-the-central-and-northeastern-united-states-and-adjacent-canada-botany-t42-l-virginicum-leaves-and-tip-of-raceme-x-23-septum-pod-and-petal-x-2-cross-section-of-seed-x-4-wild-peppeegrass-cotyledons-accumbent-and-seed-minutely-margined-pod-marginless-or-obscurely-margined-at-the-top-petals-present-except-in-some-of-the-later-flowers-a-common-weed-of-roadsides-and-waste-places-june-sept-fig-742-2-l-apitalum-willd-nearly-scentless-leaves-toothed-or-pinnatifid-coty-ledons-incumbent-as-in-PFYPT8.jpg&imgrefurl=https%3A%2F%2Fwww.alamy.com%2Fstock-photo%2Faccumbent.html&tbnid=5NzdMsR0u0dH3M&vet=12ahUKEwjwk9mA2O_0AhUS1IUKHVIyBREQMygAegQIARAy..i&docid=XOCNq507doA78M&w=1300&h=1189&itg=1&q=accumbent%20%20plants&ved=2ahUKEwjwk9mA2O_0AhUS1IUKHVIyBREQMygAegQIARAy

Achene

Achene is a term used to refer to a dry fruit with one seed. The seat coat is said to be attached to a thin, dry fruit wall, by a tiny seed stem, making it easy for the seed to be removed from the fruit. Plants that have achene fruits include buckwheat, strawberry, caraway, quinoa, amaranth, cannabis, rose plant, and the buttercup plant family. Achene fruits do not open upon maturity, like most fruits do, in order to release the seed (Waraich et al., 2011).

Achenes occur in several variations depending on the plant species. For instance, some achenes are winged in nature, like in maple trees hence they are referred to as samara. The second variation of achene is that, some have hair-like structures that allow them to sway in the wing, and are referred to as tumble fruits; common in *Anemone virginiana*.

Figure 9: An image of an achene .

Source:https://www.google.com/imgres?imgurl=https%3A%2F%2Fcdn.britannica.com%2F98%2F131098-004-65A20965%2Fgenus-Dandelion-achene-Taraxacum.jpg&imgrefurl=https%3A%2F%2Fwww.britannica.com%2Fscience%2Fachene&tbnid=MWcSZrk34XV6QM&vet=12ahUKEwjB2KrM2-_0AhVN9hoKHRzrAg8QMygAegUIARCzAQ..i&docid=KTsESUmTLkJLoM&w=550&h=412&itg=1&q=achene%20in%20plants&ved=2ahUKEwjB2KrM2-_0AhVN9hoKHRzrAg8QMygAegUIARCzAQ

Acicular

Acicular is a term used in plants to refer to plant leaves that are slender in shape. Like most living organisms, plants are known to survive in different climatic conditions. Some species are able to survive in different regions which experience different climatic conditions. Such species have such special capabilities because they are able to quickly adapt to changes, enabling them to survive even in the worst climatic conditions. Furthermore, the continued increase in changes in the world climate is pushing living organisms, especially plants, to pick on a new trend of quickly adapting to new climatic conditions to avoid ending up extinct.

Acicular leaves serve different purposes in plants on which they are found on, however, the functions are similar to those of normal leaves on other plants, with the only difference being, on the mechanism of accomplishing such tasks. One, they help plants in breathing by providing tiny stomata holes which aid in air circulation. Since acicular leaves are slender in shape, most of the stomata holes are located on the leaf edges. The second function is that acicular leaves help in water retention (Van Dijk et al., 2021). This is because, their slender shape reduces the rate at which water is lost through transpiration is slower, hence plants with acicular leaves end up retaining high amounts of water within their structure. The third purpose of acicular leaves, though it is not really a function, rather an added advantage to plants with acicular leaves is that, they remain green in all seasons. This is because; one, acicular leaves for not experience any abscission since they are much capable of retaining large amounts of water within the plant structure; and two, due to high water retention, they remain attached to the tree branch for the whole year, maintaining the green beauty of the trees.

Acicular leaves are further divided into different categories, mainly because of their shape. The first category is that of wide acicular leaves. This might be confusing, and in most cases, this category of acicular leaves is assumed to not be correctly categorized. However, wide acicular leaves have sharp edges, not only at the tips but around the entire leaf edges. The second category of acicular leaves is that of leaves that are extremely thin and long in shape. These are the most recognized acicular leaves to many.

Plants are known to have acicular leaves include the likes of pine trees which have the thin category of acicular leaves; firs which also have the long and pointed category of acicular leaves; cedar trees which also have the thin acicular leaves; and the Aesculus trees which have the wide category of acicular leaves.

Acicular leaves have several advantages and disadvantages in plants. As for the advantages; acicular leaves are good in consumption of water and Energy, for their structure reduces the rate of water loss and energy loss from the plant. Secondly, acicular leaves stay attached to the plant's branches for long retaining the beautiful natural appearance of plants. Last but not least, acicular leaves are resistant to inclement weather in that they are able to keep the plant strong and survive even in extremely cold weather, such as during snows. The drawbacks of acicular leaves are; one, they have a slow growth rate (Van Bruggen et al., 2016). The slow growth rate is contributed by the fact that their slender shapes make them have small pores; with less pores, the rate of absorption of carbon dioxide from the atmosphere is slow, hence less photosynthetic activities, leading to less food production. Furthermore, all plant organs rely on food produced through photosynthesis, for their growth. The second disadvantage of acicular leaves is that, despite being able to retain water and resist inclement weather conditions; they cannot withstand extremely high temperatures, of approximately forty degrees Celsius and above. Under such high weather conditions, they are forced to dry up and fall from the trees.

Figure 10: An image of acicular plant leaves.

Source: https://www.shutterstock.com/search/acicular+leaf

Acropetal

Acropetal is a term used to refer to the succession of leaves or flowers during development, from their bottom to the top. Acropetal development is mostly known to support the growth of young leaves above the old leaves, such that, the old leaves remain at the base, with young leaves shooting higher at the apex region. Acropetal succession is common in racemose inflorescence. Inflorescence is a term used to refer to the flower arrangement pattern in plants, which is different from one plant species to another (Van Breusegem et al., 2001). With racemose inflorescence, the main axis of the plant continues to grow, even as the flower reaches its growth limit. Racemose inflorescence supports acropetal opening in that, the internodes of the flower leaves continue being elongated as the axis of the shoot gets condensed.

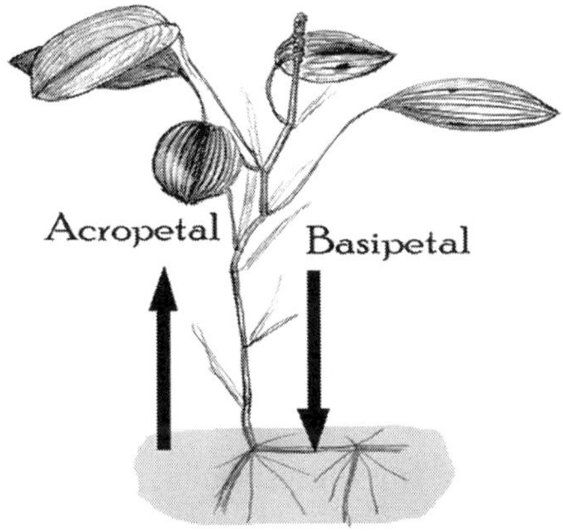

Figure 11: An image showing acropetal opening.

Source: https://www.researchgate.net/figure/Directions-of-acropetal-and-basipetal-translocation_fig6_8897675

Acrophyll

Acrophyll is a term used in plants to refer to the leaf in climbing plants, which are produced once the plant becomes mature; and their distinction from leaves produced by immature plants is clearly visible.

Figure 12: An image of Acrophylls.

Source:https://www.gabriellaplants.com/products/4-ficus-deltoidea

Actinodrous

Actinodrous is a term used to refer to a specific leaf venation where three of more veins diverge radically from a single point, and run out towards the leaf's margin. Actinodrous leaf venation is common in *Acer pseudoplatanus* plants.

The actinodrous leaf venation can either be basal or suprabasal. Basal actinodrous leaf venation has three primary veins originating from the base of the leaf, diverging radially to the margin. Suprabasal actinodrous leaf venation has the three primary veins originating in the region between the leaf base and middle of the leaf, and running towards the margins.

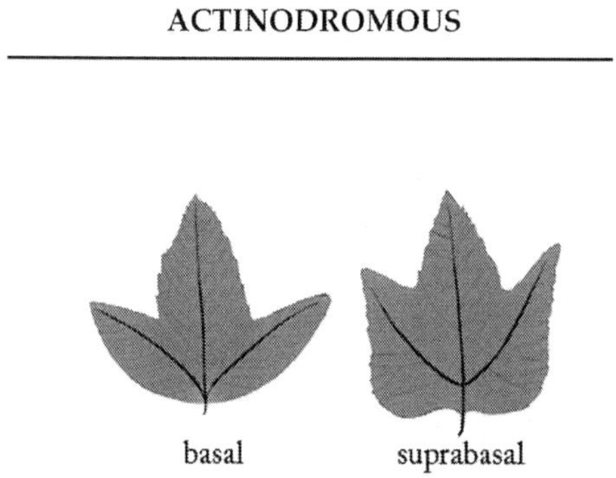

Figure 13: An image showing actinodrous leaf venation.

Source:http://sweetgum.nybg.org/science/glossary/glossary-details/?irn=3067

Actinomorphic

Actinomorphic is a term used to refer to the star shaped flowers which can also be referred to as radially symmetrical flowers or regular flowers. Actinomorphic flowers are divided in such a way that, one is able to obtain two equal halves along any plane; meaning that they have multiple lines of symmetry. The divisions could be done in such a way that, identical sections are obtained, which are related to each other based on rotation about the center of the flower. For instance, the division of an actinomorphic flower could lead to sections with only one petal.

However, not all actinomorphic flowers could be divided into equal halves. Some of them, such as the likes of Oleander flower cannot be divided into symmetrical halves in the guidance of a certain number of longitudinal planes passing through the axis.

Actinomorphic flowers exhibit a basal angiosperm character. They could also be said to be made up of clusters of tiny flowers which have been arranged into a roughly radial symmetry of the form: head, calitulum and pseudantium.

Examples of actinomorphic flowers include the likes of lily, buttercup, Trillium which has three lines of symmetry, and Salmonberry which has five lines or symmetry.

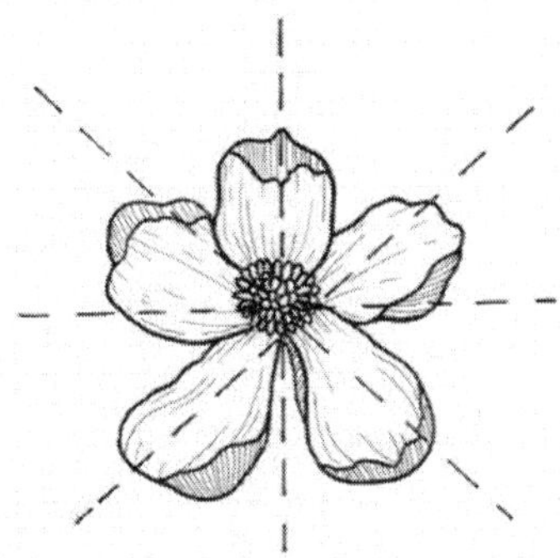

Figure 14: An image showing an Actinomorphic flower with five lines of symmetry.

Source:https://bio.libretexts.org/Bookshelves/Botany/A_Photographic_Atlas_for_Botany_%28Morrow%29/08%3A_Angiosperms/8.02%3A_Flower_Morphology

Adventitious

Adventitious is a term used to refer to plant roots that form from the non-root tissues of a plant. In some situations, adventitious roots may also be formed due to the response to stress exerted on primary roots during their normal development process. The main difference between adventitious roots and primary roots is that; primary roots could be short-lived hence they end up being supplemented by roots that grow from the stem, a habit common in orchid plants, bromeliads plants, epiphytes and basically most of the monocotyledonous plants (Van der Kooi & Ollerton, 2020). Basically, adventitious roots are mostly found in plants growing in swampy regions such as rice plants.

The modifications of adventitious roots for aerial support lead to the formation of roots referred to as prop roots which are common in corn plants and fig trees. The modification is a natural process.

In tropical raintorests, trees develop large prop roots which grow on the horizontal branches located close to the ground level. Prop roots on the rainforests trees develop for additional anchorage and support, since such trees are quite huge hence in need of additional support. Bulbous plants, which are also found in rainforests do develop contractile adventitious roots, which, as they offer support to the tree, they continue pulling the bulb deeper into the ground. The adventitious roots which develop from the stem are referred to as stilt roots, such as the ones found in mangroves.

Some climbing trees do also develop adventitious trees in their later stages of development. This is because, as they continue to increase in height, they get weaker hence the need for additional support.

Modified stems in plants, which end up turning into adventitious roots are categorized into four. The first type of modified stems is referred to as the Stolon. The stolons are roots that grow along the ground, more of being in parallel with the ground surface. They can be found in strawberries. The second category is the Rhizomes (Twidwell et al., 2013). Rhizomes are similar to stolons with the main difference being that rhizomes grow under surface, in parallel alignment with the ground. The third category is the Tubers, which are a modification of rhizomes in the sense that they both grow underground but they do swell, unlike rhizomes. The fourth category is the tendrils, which are found on climbing plants.

Adventitious roots could either be grown solely for support purposes, however, there are some plant species that later modify their roots to become points of food storage such as in carrots, fleshy tuberous plants among others.

There exist different rules which govern the selection of suitable non-root tissues to be used in the formation of adventitious organs. The first rule is the extent of differentiation. The not-fully developed cells are the ones most preferred when it comes to the development of adventitious roots because they regenerate easily compare to cells from fully developed cells. The second rule is that of the Ontogenetic age. Young cells are believed to regenerate at a faster rate compared to old cells. The third rule is based on Physiological age. Recent cells present a higher performance in terms of cell regeneration compared to old cells. The last factor to be considered is on Dormancy state. Cells that have been in a dormant state for long regenerate at a slow rate.

The formation of adventitious roots is a step-by-step process that follows through four different phases. The first phase is that of induction. As the name suggests, induction is the stage where the development of the adventitious roots is commenced through the preparation of development materials. It takes a period of twenty-four hours for it to be complete. The second stage is referred to as Initiation (Taiz et al., 2015). In this stage, the development materials pick up the growth process. It takes a period of two to four days for it to be complete. The third phase is that of the formation of root primordia that takes a period of four to five days. The last phase of adventitious root development is referred to as root elongation, which as the name suggests, allows the roots to fully develop; and it takes a period of five to six days.

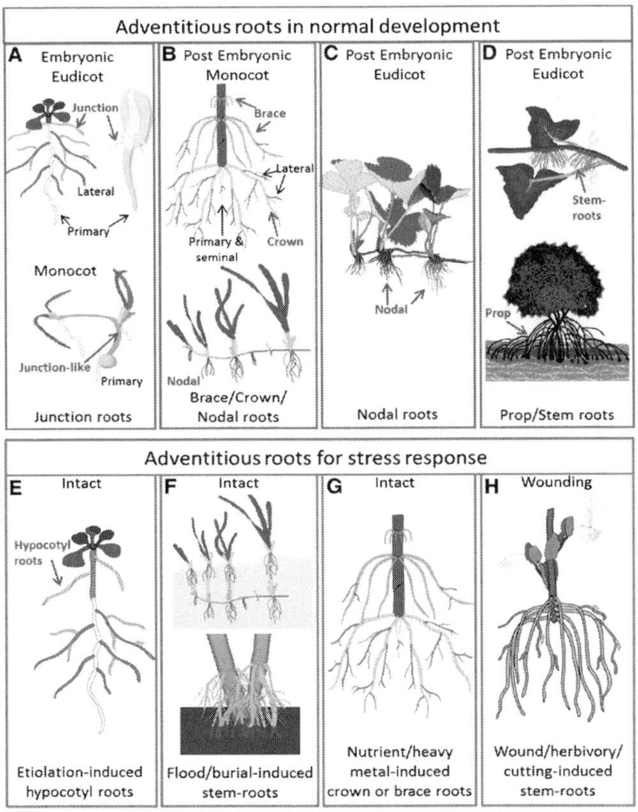

Figure 15: An illustration of Adventitious roots development.

Source: https://academic.oup.com/plphys/article/170/2/603/6114063

Adnate

Adnate is a term used to refer to the fusion of plant organs that are totally different from each other. This union of dissimilar organs is not based on any principles, rather, the organs to be fused are selected at random by nature. In flowers, adnate is used to refer to the fusion of two or more flower organs. For instance, stamens and petals could be fused together. A good example of Adnation is in the flowers of the *Primula vulgaris* plant whereby the stamens are fused to the top be of the pistil (Szpak, 2014).

The fusion of different organs, more so in flowers could be along the axis. Axis is a term used in plants to refer to a reference point which could be used to identify the center region of an organ. However, in some cases, Adnation could be done along any desired region as desired by Mother Nature.

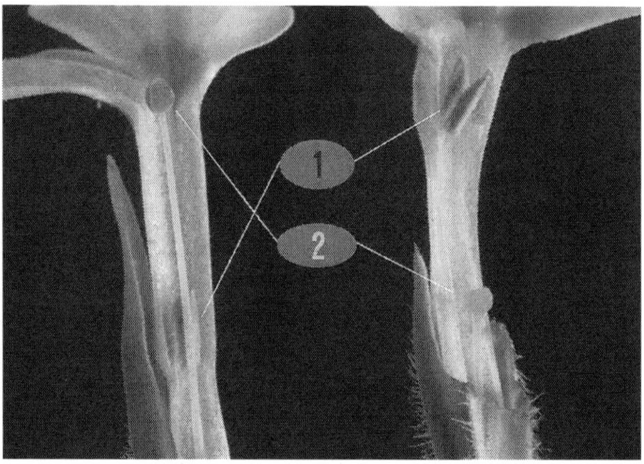

Figure 16: An illustration of Adnate organs.

Source: https://www.google.com/amp/s/www.differencebetween.com/difference-between-adnate-and-connate/amp/

B

Bark

Bark is the term used to refer to the outermost layer of roots and stems of woody plants. Some of the plants known to have bark are shrubs, woody vines and trees. The tissues outside the vascular cambium make up the bark in plants. The term bark is not a technical term. Bark is composed of the inner bark and the outer bark and it overlays the wood. In older stems, the inner bark is made of living tissue. It is composed of the innermost layer of the periderm. Different from the inner bark, the outer bark is made of dead tissue and is found at the surface of the stems. Outer bark is also found along with parts of the outer periderm as well as all the tissues found on the outside of the periderm (Stenberg, 2017). The rhytidome is the outer bark on trees that lies external to the living periderm. Some of the products are derived from bark. There are a number of products made from bark. They include spices, wall coverings, bark shingle siding and flavorings. Tanbark is used in medicines, cork, hallucinogenic chemicals, poisons, latex, resin and tannin. Bark has been used in making ropes, canoes and clothes. It is at times used as a surface for map making and paintings. There are certain plants grown for their interesting and attractive bark decorations. Bark is at times used as a landscape mulch.

Figure 17: There are both inner barks and outer barks.

Source: https://www.newworldencyclopedia.org/entry/Bark

Basipetal

Basipetal is a form of growth in flowers and plants. In flowers, it is a type of flower arrangement in the floral axis where by the main axis of the plant terminates in a flower whole the buds and newly formed flowers arise laterally from the bottom parts of the plant. Basipetal is the opposite of acropetal (Singh, 2015). In Basipetal arrangement, one could say that new flowers are arranged at the bottom while old flowers are arranged at the top. Basipetal arrangement is also known as the cymose inflorescence. This kind of arrangement is mostly noted in plants like jasmine and clerodendrum. Acropetal arrangement is noted in plants like Gulmohar and Caesalpinia.

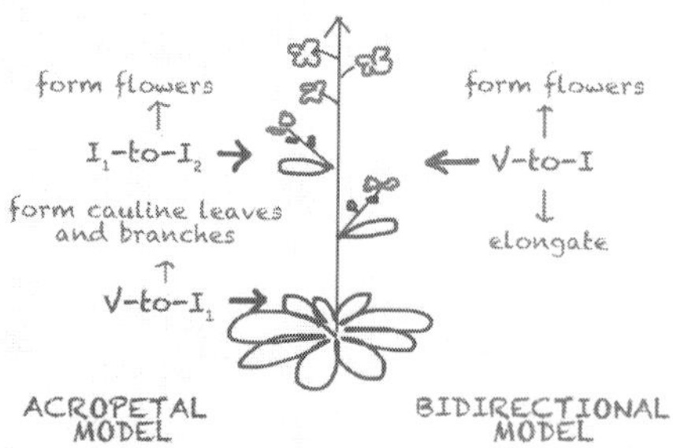

Figure 18: Basipetal is an arrangement in both leaves and flowers.

Source: https://floweringhighlights.org/2014/08/12/flowering-highlights-on-gibberellins-and-life-decisions/

Bathyphyll

Bathyphyll is a term used to refer to leaves in climbing plants that are produced by immature plants. It could also refer to the entire climbing plant. These leaves are very unique compared to other leaves produced by climbing plants. They are also very different from leaves produced by the same plant when they mature. The uniqueness may be due to the fact that they are not specialized like other leaves. It is also green like other plants.

Figure 19: Bathyphyll are leaves produced by immature climbing plants.

Source: https://www.healthbenefitstimes.com/glossary/bathyphyll/

Berry

Figure 20: Wild berries are produced by wild plants.

Source: https://www.healthline.com/nutrition/11-reasons-to-eat-berries

Berries are fruits of plants. They are small, pulpy and edible. In most cases, they are brightly colored, rounded and very juicy. Their taste may be tart, sour or sweet. In some cases, the berries may have seeds or pips. There are different types of berries namely blackcurrants, white currants, red currants, blackberries, raspberries and strawberries. In Britain, they are classified under soft fruits which is a term used to refer to such fruits. It is said that the term berry differs from the botanical or scientific definition of a fruit produced by the ovary of a single flower (Schuurink et al., 2006). After development, the ovary wall develops and becomes the pericarp which is the edible part. Blackberries, raspberries and strawberries are examples of berries but are not included in the botanical definition. They are considered multiple fruits composed of mulberries and aggregate fruits. Examples of giant berries include pumpkins and watermelons and are categorized under pepos. The term baccate or bacciferous is used to refer to plants that bear berries.

Biennia

Figure 21: Biennia refers to plants that take two years to develop.

Source: https://www.thespruce.com/what-is-a-biennial-plant-4134320

The term biennia is used to refer to flowering plants that take about two years to complete a biological life cycle. The first year is characterized by the plant undergoing primary growth. During this development, the roots, stems and leaves develop. This is characterized by the development of vegetative structures. In this period the stem of the plant is short while the leaves are low to the ground. This results in the formation of a rosette. In the early stages of the second year the plant enters a period of dormancy during cold months. Before flowering biennia may need vernalization or cold treatment. During summer of spring, stems of these plants greatly elongate. It is followed by the flowering of the plant resulting in the production of seeds and fruits. The final stage is the plant dying.

Bifoliate

Bifoliate is a term used to refer to plants with two leaflets. For palmately compound leaves, the two leaves are attached at a common point. The point of the attachment is at times called the node. In single plants, the leaves may be arranged parallel or alternate to each other. In palmately compound leaves, the point of attachment tends to be very strong to support both leaves. Some palmately compound leaves Bifoliate in nature include the Bauhinia yunnanensis also called the butterfly tree or kanher.

Bilabiate

Bilabiate is a term used to refer to flowers with two lips in a bilabiate corolla. The bilabiate lips include the upper corolla and the lower corolla. It is also called the two-lipped corolla. They play a vital role in attracting bees and other insects for fertilization. Bilabiate flowers are a strong indication of parallel evolution. It is a form of adaptation in flowers to limit the amount of pollen taken by insects more so bees. About 38 angiosperm families have been said to have bilabiate tendencies. They tend to have dorsal pollen transfer and dorsiventral organization. They have a part called the walk around flower.

Binomial nomenclature

Binomial nomenclature is also called the double naming system. It is a formal system used in the naming of various species of living things. In binomial nomenclature, a living thing has two names. Both names use Latin grammatical forms. The first part of the name is the generic name. The generic name is used to identify the genus to which the species belong. The second name is the specific name that identifies the species to which the animal belongs. Various internationally agreed codes of rules are used when applying the binomial nomenclature (Smith & Knapp, 2001). There also some principles that are followed when applying the binomial nomenclature. When using the system, the first name is always capitalized in writing. The second name is not capitalized. Both the generic and specific names are italicized in normal text. This system has been very useful in identifying many species of plants allowing their existence to be documented.

Bipinnate

Bipinnate is a term used to refer to plants whose leaflets have further been subdivided in a pinnate arrangement. This form of leaflet arrangement is most common in trees. In most cases, the leaves of the trees are doubly compound while the leaflets have smooth margins. Petioles with these leaflets appear on the rachis. They are further subdivided in the secondary rachises. Pinnule is another botanical name for Bipinnate (Simon & Petrášek, 2011). Pinnule describes leaflets that have been further pinnately divided. The honey locust is a common North American tree species with bipinnate leaves. Other examples include Jerusalem thorns, chinaberries, flamegolds, silk trees and bailey acaiacs. As it is easy to confuse between tripinnate leaflets and bipinnate leaflets, it is very important to note whether the leaflets attach to the first or second rachis. For attachment in the first rachis then the plant is bipinnate while the latter is tripinnate.

Bipinnatisect

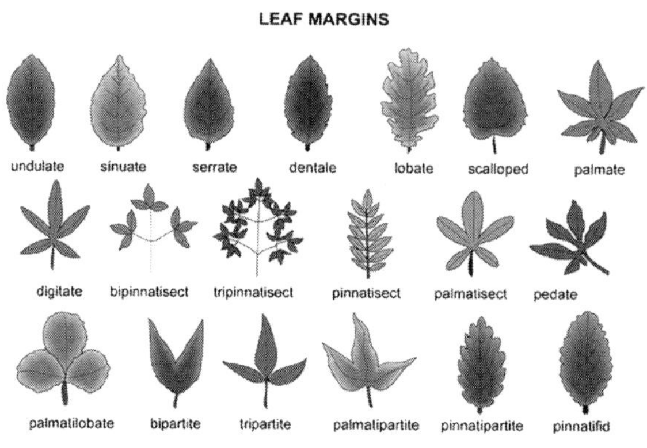

Figure 22: Bipinnatisect describes the appearance of a leaf's margin.

Source: https://infovisual.info/en/biology-vegetal/leaf-margins

4Bipinnatisect is a term used to refer to plants with two levels of petioles which segments are sessile. Pinnation refers to the arrangement of leaves. Petioles form the pipeline through which products of photosynthesis are moved from the leaves to the rest of the plant. This form of arrangement has been useful in classifying plants. If a pinnatisect has leaves that have deeply dissected segments, they are classified as Bipinnatisect. The term Bipinnatisect is used when describing the morphology of the plant.

Bisexual

Figure 23: Bisexual plants have both male and female reproductive organs.

Source: https://www.quora.com/What-is-the-difference-between-monecious-and-bisexual-plants

The term bisexual is used when referring to flowers with both female and male reproductive organs within the flower. This means that the flower has both carpels and stamens are found in that flower. Bisexual flowers are different from unisexual flowers in that unisexual flowers have the male and female organs in the same plant but in different plants. Monoecious plants are known to bear bisexual plants. Examples of bisexual flowers include mango, country bean, Chile, long bean, tomato, Hibiscus, Brinjal, Mustard, Daffodil, sunflower, rose and Lily. Bisexual flowers are categorized under self-pollinated flowers as they do not rely on wind or insects for pollination. Bisexual flowers are at times called the perfect flower (Rusman et al., 2019). By having both reproductive organs within the same flower, reproduction easily occurs in the flower. The structure of the flower is as follows: the flower has a carpel, stamen, corolla and calyx. The sepal is the unit of the calyx. The sepals are green in color. The corolla is located above the calyx. The corolla is the unit of the petal. The male organs contain both the anthers and the filament while the female organs contain the ovary, stigma and style. Bisexual flowers may be complete or incomplete.

Bloom

Bloom is a term used to refer to a variety of things with regard to plants. It describes the blooming stage of flowering plants, the appearance of blossoming flowers, the actual bloom or the waxy outer coat of some plants. When referring to the blooming stage of flowering plants, plants tend to bloom after the vegetative stage. This stage is also referred to as the flowering stage. During this stage, plants are very delicate. The breaking of flower petals during snow is used to indicate the arrival of spring. This is a very complex genetic process in the floral façade. A gene called Apetala1 tells flowers when to bloom. It is a lone master gene. It also triggers the reproductive development of the plant thereby telling the plant when to blossom. When the gene, Apetala1 is active, the plant is blooming. If the gene is in active in a plant, it may tend to have very few flowers and leafy shoots growing in place of blossoms.

Bole

The bole also called the trunk is the stem of a tree. It also acts as the main wooden axis of the tree. The bole plays a vital role in the identification of a tree. Depending on the species of the tree, there are different markings from the bottom to the top of the trunk. The bole plays a vital role in trees during timber production. Both woody and non-woody plants can have boles. Examples of non-woody plants include monocots and palms (Reid et al., 2009). The internal physiology of the bole in both woody and non-woody plants may be different. A common tendency in all plants is that the trunk tends to thicken gradually due to the formation of secondary growth. Boles are very vulnerable, any long duration exposure to sunlight results in sun burns. The part of boles cut down during logging are called logs. The part of the bole remaining in the ground after the tree has been felled is called the stump.

Bract

Bract is a term used in botany to refer to specialized or modified leaves more so those with reproductive structures such as cone scale, inflorescence and flower. Foliage leaves and bracts are very different in most cases. Brats may be small, large or of a different texture, shape or color. Bracts are very different from certain parts of the flower. The term bracteolate or braceate is used to refer to plants with bracts. A plant without bracts is called ebracteolate or ebracteate.

Branchlet

Figure 24: A Branchlet is a subdivision of a branch.

Source: https://www.collinsdictionary.com/dictionary/english/branchlet

Branchlet is a term used to refer to the sub-division a branch. It is also called a small branch. There are certain species of plants with branchlets while in others the leaves are directly attached to the main branch. Branchlet is normally thinner compared to the normal main branch and is less strong compared to the main branch. Branchlets connect the leaves to the main branch. In this sense, it allows transportation of water and mineral salts to the leaves. They are also actively involved in photosynthesis. In some plants, the branchlets are pruned to allow more energy into the branches. Branchlets are also connected to the nerves of the plant. They help support the plant (Poff & Zimmerman, 2010).

Brevideciduous

Brevideciduous is a term used to refer to plants that lose all their leaves briefly before growing new ones. The plant, therefore, becomes leafless for a given duration of time. In most cases, it takes about two weeks. This often happens when the old leaves have to be removed from the plant for new leaves to grow. This normally occurs in sub-tropical woody species and tropical species. The loss of leaves may occur during the winter or autumn. In some cases, leaves are lost during the dry season. These plants may mimic behaviors of deciduous plants that lose their leaves in cold seasons and severe drought. The loss of leaves mostly occurs before the plants begin to flower. For instance in kowhai plants, they lose their leaves at the beginning of spring. It is followed by a phase of heavy flowering. The flowers then fall off allowing the growth of new leaves.

Brochidodromous

Brochidodromous is a term used to refer to a kind of leaf venation. Brochidodromous is used to refer to leaves with second order veins. It is a pinnate leaf venation characterized by secondary veins joining as series or arches. This means that the leaves do not terminate at the margins. These arches may form loops. For loop veined plants, the main veins emerge from the mid rid at regular intervals and at the margin they turn upwards towards the apex and looping to join the next vein upwards. This kind of venation greatly influences the leaf's performance. Although we know that for Brochidodromous plants, their veins form loops, there are differences in the loopiness. This means that different plants have different orders of interconnections. For plants with higher-order loopiness, this adaptation protects them from herbivory (Pauchard et al., 2009). Depending on the shape of the Brochidodromous, the loops in leaves will vary. Due to their topologies, Brochidodromous plants will record a higher leaf mass per unit area. Compared to other leaves, they will have a greater lifespan. This contributes to the abundance of evergreen plant species. There is a specific biomechanical support used in Brochidodromous plants. This leaf venation contributes to the branching support.

Brochus

Brochus is the term used to refer to the width of one lumen of pollen grain reticulum and a half the width of surrounding muri. Its plural is brochi. The muri is a wall structure that encloses the space. This brings about the term homobrochate and heterobrochate. In homobrochate, the lumina is of the same size while in heterobrochate the lumina has different sizes. These terms are used to define the morphology of pollen.

Bryophyte

Figure 25: Bryophyte is a seedless plant.

Source: https://www.biology-questions-and-answers.com/moss.html

Bryophyte is a term used to refer to non-vascular seedless plants. Examples of bryophytes are liverworts, hornworts and mosses. All these plants are of different divisions. A common characteristic in most bryophytes is that they do not have complex tissue organization. However, these plants demonstrate diversity in ecology and form. Bryophytes are widely distributed throughout the world. Compared to seed-bearing plants, bryophytes are small in size. They also demonstrate an alternation of generations between the dependent sporophyte generation that produces the spores and the independent gametophyte generation that produces eggs, sperms and sex organs (Prajapati & Modi, 2012). Bryophyte sporophytes in most cases do not have a complex vascular system. They, therefore, produce one spore-containing organ. For bryophytes, the gametophyte generation is perennial. They are also photosynthetically independent of the sporophyte. This results in the formation of an intimate interconnection with the gametophytic tissue. This is mostly noted at the foot or base of the sporophyte. For bryophytes, the gametophyte is the conspicuous and long-lived generation. Gametophores of bryophytes have structures resembling leaves, roots and stems. These structures can also be found in sporophytes of vascular plants. Spores are released by sporophytes. Gametophytes develop from thee spores.

Bulb

Figure 26: Bulbs are mostly found on the roots of plants.

Source: https://www.gardeningknowhow.com/garden-how-to/gardening-by-zone/zone-8/zone-8-bulbs.htm

Bulb is a term used to refer to a modified stem. These kinds of stems occur at the resting stage of certain seed plants. In most cases, bulbs are large, globe-shaped and have membranes. The underground bulb membranes are composed of freshly overlapping leaves. In some plant species, bulbs are expanded leaf bases. The bulb's fleshy leaves are at times used by plants as food reserves. This enables the plant to remain dominant when there is insufficient water. After the dominant period, the plant is able to resume the initial active growth during prevailing conditions. Bulbs can be grouped into two main groups. The onion typifies one type. In this group, the bulb has a thin paper covering that offers protection for the fleshy leaves (Neumann et al., 2009). The scaly bulb typifies the second group. It is mostly noted in true lilies. They have naked storage leaves and are unprotected by any paper covering. There are different bulb sizes with some being pea-sized while others being large-sized. In some cases, an individual bulb may weigh more than 15 pounds. Hyacinths, tulips and narcissus are able to rapidly produce flowers in the presence of bulbs. Lilies produce bulbs and they flower in the summer. Meadow saffron produces bulbs but flowers in the fall. Both the amaryllis and lily families are referred to as bulb-producing species.

Bulbel

Figure 27: A bulbel is a separable bulb formed in the flowering plant.

Source: https://educalingo.com/en/dic-en/bulbel

Bulbel is a separable bulb formed in some plants. The outer layers of the bulbs are covered with dead tissue to protect the inner bulb. Onions are good examples of bulbels.

Bulbil

Figure 28: Bulbils are plants reproduced vegetatively from auxiliary buds.

Source: https://en.wikipedia.org/wiki/Bulbil

Bulbils are small young plants vegetatively reproduced from axillary buds. The auxiliary buds are on the parent's plant stem. The young plant can also be reproduced in place of a flower on an inflorescence. The cones found on plants are the produced bulbils. Bulbils have identical genetic material with their parents. Bulbils are formed through asexual reproduction. In some cases, the bulbil will form a new stand-alone plant. Certain bulbils are said to meet the botanical criterion to be considered as true bulbs. There are some bulbils that are not considered to be true bulbs. There are a variety of different morphological forms of bulbils: some are considered to be true bulbs while others are not. There is a distinction between bulbils and bulbs (Nikunj et al., 2014). Plants like lilies and onions are bulbous plant groups. They produce bulbils in the form of a secondary small bulb. Bulbils produced by onions and lilies meet the botanical criterion to be labeled as true bulbs. For this reason, bulbils produced by bulbous plants are considered bulbs. There are bulbils that are not considered as bulbs. They include bulbils produced by plants under the non-bulbous plan groups. Most of these plants fall under the subfamily Agavoideae. Though they produce bulbils, their bulbils are not considered as true bulbs.

Bulblet

Figure 29: A fern bulblets.

Source: https://www.flickr.com/photos/anitagould/3037269344

Bulblets are small bulbs or bulb-shaped bodies that are borne above the ground. They are mostly found on the stem of a leaf axil. Bulblets are produced through asexual reproduction. Bulblets tend to grow on bulbs or attached to bulbs of plants. There are fern and onion bulblets among others. In onions, the bulblets are visible in onion trees. The formation of bulblets is also common in garlic. Bulblets in onions are at times referred to as top onions. Among the species of fern, there are those called the fern bulblets. They include *Cystopteris bulbifera*. This species of fern are low-growing rock ferns characterized by creeping stems and narrow elongated deltate fronds (Novello et al., 2020). They produce bulblets along the bottom of the fronds. This feature makes this species of fern to be easily identified. Ferns have sori on their leaflets and at the bottom, there are green bulblets. The bulblets may be found at the base of some leaflets' undersides. Bulblets produced by ferns are globoid in shape and are cleft at the top. Eventually, the bulblets become larger than the sori and when they mature they drop from the leaves. The fallen bulblets during favorable conditions develop into small ferns. Bulblets therefore play a crucial role in the reproduction of ferns.

Bullate

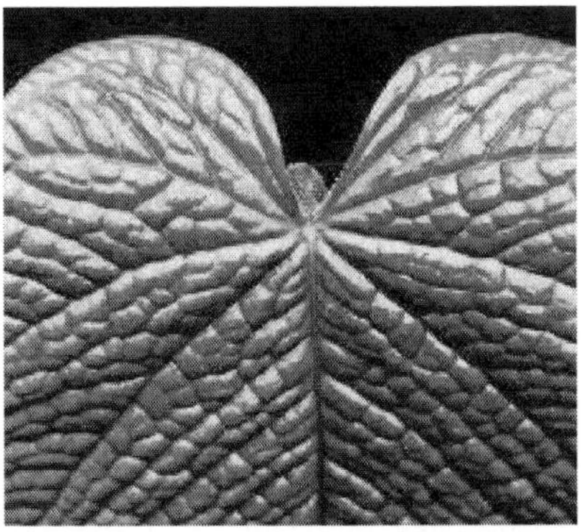

Figure 30: Bullate describes the appearance of leaves.=

Source: https://web.facebook.com/speakinplants/photos/bullate-leaves-surface-its-a-sign-that-the-specie-most-likely-inhabits-relativel/3402758259742934/?_rdc=1&_rdr

Bullate is a term used to refer to the appearance of leaves in plants. Bullate leaves are said to have semi-globular swelling. A wide range of plant families have Bullate leaves. This includes plants in familiar temperate species and tropical ornamentals. Bullate leaves are a form of adaptation in plants. Among the advantages is the fact that it increases the leaf's ability to capture light from all directions. The Bullate shape increases the leaf's total surface area hence the large amounts of sunlight capture. Another advantage is that it repels water allowing the quick drying of a leaf surface by use of channeled drainage. This adaptation is very suitable for plants that grow in low light and very wet areas.

Burr

Burr is a dry fruit or seed having teeth or hooks. The presence of burr in plants is very important as it allows the spread of seeds of the bur plant. Epizoochory is the process in which the seeds are spread. The burr gas hooks are useful in catching on. The hook can get attached to fabric or fur. As the burr contains the seeds, when they get attached they are transported with the attached hook. The hooks or spines are also useful in giving plants protection against herbivores. Burr plants are often classified as bothersome due to their ability to stick. In some cases, plants do not have burrs but have some features that resemble burrs (Mercuri et al., 2002). For instance, cactus have thorns which may function as burrs. Xanthium species are examples of burr-bearing plants. When growing in dense groups they are single-stemmed but when growing singly they spread. The size and shape of plants will influence the number of burrs in a plant. Other examples of burr-bearing plants include cocklebur, puncturevine, American lopseed, Clayton's sweetroot, herb Bennet, white avens, yellow avens, cleavers and soft agrimony.

Buttress root

Figure 31: Buttress roots are common in trees with shallow roots

Source: https://en.wikipedia.org/wiki/Buttress_root

Buttress roots are a type of roots characterized by large and wide roots on every side of the tree. Plants with Buttress roots are mostly found growing in soil having poor nutrient levels. The extensive roots help anchor the tree while allowing it to gather all the needed nutrients. The buttress roots are tension elements. They occur mostly on the side away from stress. The main roots of other plants may inter-wind with Buttress roots resulting in the creation of an intricate mesh (Matyjaszczyk & Śmiechowska, 2019). This mesh may help support the trees surrounding it. If the buttress roots are spread out horizontally, they cover a larger area and are able to collect more nutrients. Most buttress roots are found on the upper soil layer. This is because the main nutrients are found in the upper soil layer.

Byssoid

Byssoid is a term mostly used to refer to lichen with wispy growth forms. This kind of lichen has a teased wood appearance. An example of a byssoid lichen is *Coenogonium implexum*. There are two main types of byssoid lichens. In the first type, the cottony structure is dominated by photobiont filaments while in the second type the cottony form is made up of fungal hyphae. The hyphae in this case contain a typically coccoid photobiont.

C

Caducous

The splitting of a mature plant structure along a built-in line of weakness in order to liberate its contents is known as dehiscence. This is typical of fruits, anthers, and sporangia. This can sometimes include the total detachment of a par; structures that open in this manner are referred to as dehiscent. Structures that do not open in this manner are referred to be indehiscent, and their contents are released through other mechanisms such as disintegration or predation (Mandel, 2019).

Some flower buds (e.g., Platycodon, Fuchsia) undergo a similar process to dehiscence, however this is rarely referred to as dehiscence unless circumscissile dehiscence is included; anthesis is the common term for flower opening. Dehiscence may or may not involve the loss of a structure due to the abscission process. The structures that have disintegrated are said to be caducous.

Caespitose

Growing in dense clumps or tufts; e.g., the growth form of some species of Poaceae and Cyperaceae. The clumps are also called tussocks.

Tufted or turf-like, e.g. the growth form of some grasses.

Calcarate

Having a spur; spurred.

Spur: A tubular projection of tissue on various plant organs. The term is most commonly used in botany to refer to nectar spurs on flowers (Lindström et al., 2013).

Nectar spurs, which can protrude from a flower's sepals, petals, or hypanthium, contain nectaries, which are glands that exude nectar.

Calcareous

Calcareous is an adjective that means "primarily or partly made of calcium carbonate," which means "containing lime" or "chalky." The phrase is used in a wide range of scientific fields.

Calcareous grassland (also known as alkaline grassland) is a type of habitat that grows on chalk and limestone with thin basic soil. Calcareous grassland plants, which include grasses and herbs like clover, are often short and resilient. Calcareous grassland is a critical habitat for insects, particularly butterflies, and is kept at a plagioclimax by grazing animals, primarily sheep and occasionally cattle. Rabbits used to play a role, but their numbers have declined due to the spread of myxomatosis, and they no longer have much of a grazing impact.

Northwestern Europe is rich in calcareous grassland, particularly in southern England's Salisbury Plain and the North and South Downs.

Figure 32: Ranscombe Farm, Medway on the North Downs. In June, these meadows are covered with chalk grassland flowers.

Source: https://en.wikipedia.org/wiki/Calcareous_grassland#/media/File:DownsRanscombeFieldToMway0734c.JPG

Callose

Callose is a carbohydrate found in plants. Polysaccharides, also known as polycarbohydrates, are the most common type of carbohydrate present in food. They are polymeric long-chain carbohydrates made up of monosaccharide units linked together by glycosidic bonds. This carbohydrate can react with water (hydrolysis) with amylase enzymes acting as a catalyst, resulting in component sugars (monosaccharides, or oligosaccharides). They have a variety of structures ranging from linear to heavily branching (Luan et al., 2020). It is produced by the glucan synthase-like gene (GLS) in multiple locations throughout a plant. It is formed in response to stimuli such as stress or damage to serve as a temporary cell wall. Callose is a -glucan, which is made up of glucose residues connected together by -1,3-linkages. It is assumed that callose synthases produce it at the cell wall and that -1,3-glucanases destroy it. Callose is critical for the permeability of plasmodesmata (Pd) in plants; plasmodesmata callose regulates plant permeability (PDC).

Calycophyll

A leaf-like structure generated by a sepal or calyx lobe that enlarges, generally many-fold, before or after anthesis, especially when the majority of the sepals or calyx lobes remain the same size.

Calycophyllum is a genus of flowering plants in the Rubiaceae family. Augustin Pyramus de Candolle described it in 1830. Mexico, Central America, South America, and the West Indies are all home to the genus.

Calyx Tube

The calyx is a leaf-like component located at the bottom of many flower species that shields the bloom while it develops. The leaf-like features are known as sepals when viewed individually. The number of sepals is frequently as much as the number of petals. While most calyces are green, there are several cases where the calyx is the same hue as the flower's petals or a completely different hue.

The calyx is required for the plant to blossom. As the flower grows, it closes firmly into a bud. The sepals are the flower's outer covering as it develops, and they are the only thing observable when the flower is still in immature form. It protects the developing blossom from the environment and prevents it from drying out (Lambers et al., 2008).

The calyx is the first portion of a flower to form and is directly linked to the plant's stem. Petals, like the rest of the flower, form inside the calyx. When the flower opens, the calyx sepals are commonly placed in an alternate pattern with the petals. It may be simpler to consider the tightly closed calyx as a nursery for the developing flower. The flower is securely wrapped in the calyx until it has finished developing and is ready to open.

Figure 33: Example of calyx.

Source: https://study.com/academy/lesson/calyx-in-flowers-definition-form-quiz.html

Many plants no longer need their calyx after flowering, and it begins to wither away; however, in certain flowers, the calyx becomes thorny or hairy and stays as a means of protection against insects that would climb up the plant.

Cambium

Cambium, also known as Cambia, is a layer of actively dividing cells in plants that exists between xylem (wood) and phloem (bast) tissues and is essential for stem and root secondary growth. The cambium is theoretically a single layer of cells termed initial cells; however, it is difficult to distinguish the beginnings from their still-undifferentiated descendent cells in practice, and the cambium, or cambial zone, is commonly referred to as a collection of cell layers. Cambial cells divide to produce secondary xylem cells along the central axis of the stem and secondary phloem cells toward the periphery of the stem. Undifferentiated cells that have retained their embryonic ability to grow and differentiate constitute the cambium. A cambium can also form within callus tissues, which are collections of cells that grow over the wounded surface of a wound to promote healing.

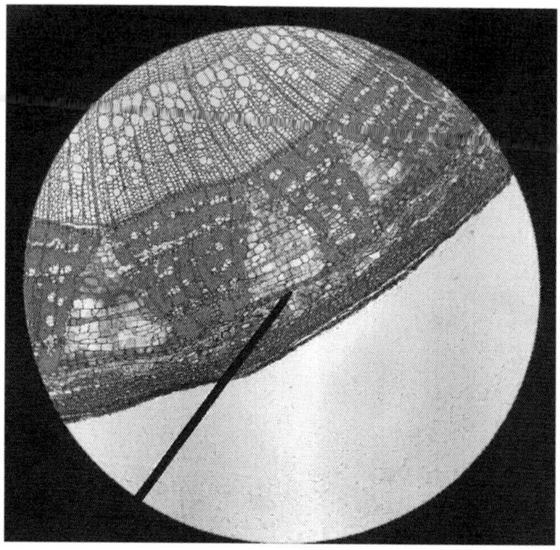

Figure 34: Cork cambium.

Source: https://en.wikipedia.org/wiki/Cambium#/media/File:Cork_cambium_1.jpg

Camptodromous

Pinnate venation occurs when secondary veins curve toward the edges, becoming nearly parallel with them in certain circumstances but not merging with other veins to form loops.

Major veins extend near to the periphery but bend before intersecting with it.

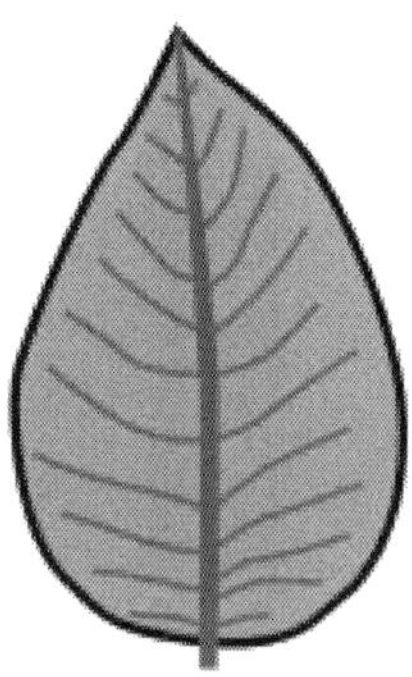

Figure 35: Camptodromous.

Source: https://en.wikipedia.org/wiki/Leaf

Campylotropous

The ovule is a female reproductive organ. It is the portion of the plant where female reproductive cells are produced and stored, and these cells eventually grow into seeds after fertilization. These seeds eventually mature and develop into adult plants. Ovules are housed in ovaries at the very bottom of a vase-like structure known as a carpel, which contains a neck called a style and an aperture at the top known as a stigma (Klassen, 2001).

A plant has six different types of ovules, which are distinguished by their forms.

The Campylotropous ovule's shape is bent, and the alignment that happens between the chalaza and the micropyle ceases to exist; as a result, the embryo sac becomes somewhat curved.

Canopy

In biology, the canopy is the terrestrial portion of a plant crop or crop produced by a group of independent plant crowns.

The canopy is the upper layer or habitat zone formed by mature tree crowns that includes other biological organisms in forest ecology (epiphytes, lianas, arboreal animals, etc.).

The phrase canopy is sometimes used to describe the outer covering of leaves of a single tree or a group of trees. The dense canopy of shade trees often inhibits light from reaching lower-growing plants (Katova, 2007).

Canopy structure refers to the organization or spatial arrangement (three-dimensional geometry) of a plant canopy. The leaf area index (LAI), also known as leaf area per unit ground area, is an important statistic for analyzing and comparing plant canopies.

The canopy looms over the understory. 90% of the animals in the rainforest live in the canopy. They cover vast areas and appear to be unbroken when observed from an airplane. Despite the fact that their branches overlap, rainforest canopy trees seldom come into contact with others. They are typically only a few feet apart.

Capillary

The movement of liquid along the surface of a solid caused by the pull of the liquid's molecules to the solid's molecules is known as capillary action.

Plants use capillary action to transfer water up the roots and stems to the rest of the plant. Water molecules (the liquid) are pulled to the stem's inner molecules (the solid). This attraction helps to force water up from the earth and distribute it throughout the plant.

Three forces govern the capillary action process.

Adhesion is the process of joining one item to another. Water's capacity to stick to the organic tissues of plants is referred to as adhesion. The capacity of molecules of the same material to stick together is referred to as cohesion. In plants, cohesion maintains water molecules together (Jeger et al., 2007).

Surface tension is the result of intermolecular interaction, which allows liquids to form a top or outer layer that acts similarly to a thin film. Surface tension is essential for the form of water drops as well as holding plant structures together as they absorb water.

Capitate

In botany, head-shaped, or collected in a head, as a dense terminal cluster of sessile or nearly sessile flowers; having a rounded head: as, a capitate stigma. In ornithology, having an enlarged extremity: as, the capitate feather of a peacock's tail.

Forming a dense, head-like cluster, such as the inflorescences of composites.

Capitulum

A capitulum (short dense spike) is a short compact spike with flowers growing straight on a broad, flat peduncle, giving the inflorescence the impression of a single bloom, as in dandelion (Taraxacum).

A pseudanthium is a flower-like inflorescence (Greek for "false flower"). Structures that are neither true flowers nor true inflorescences are sometimes referred to by this word. Flower heads, composite flowers, and capitula are all types of pseudanthia, which are inflorescences in which everything from a small cluster to hundreds or even thousands of flowers are grouped together to form a single flower-like structure. Pseudanthia can manifest itself in a variety of ways. The actual flowers (the florets) are normally small and severely reduced, although the pseudanthium itself can occasionally be rather massive (as in the heads of some varieties of sunflower) (Hirschberg, 2001).

Capsule

A capsule is a form of simple, dry, but seldom meaty dehiscent fruit produced by numerous angiosperm species (flowering plants).

A complex (multicarpeled) ovary produces the capsule. A capsule is a structure that is made up of two or more carpels. The term locule (or cell) refers to a chamber within the fruit of (flowering plants). Fruit is classed as uni-locular (unilocular), bi-locular, tri-locular, or multi-locular based on the number of locules in the ovary. A gynoecium's locule count may be equal to or less than the number of carpels. Septa separate the locules, which contain the ovules or seeds.

In most cases, the capsule is dehiscent, which means that when it matures, it splits apart (dehisces) to release the seeds inside. Some capsules, such as those of *Adansonia digitata*, Alphitonia, and Merciera, are indehiscent.

Capsules are often classified into four types, depending on the type and location of dehiscence

- Loculicidal capsules
- Septicidal capsules
- Circumscissile capsules
- Poricidal capsules

Carinal Canal

A longitudinal canal in the stem internode of Equisetum (horsetail) and some of its fossil ancestors, located radially opposite a high stem ridge (carina). Each carinal canal is situated on the inside of a vascular bundle. The carinal canals are thought to function in water conduction and are thought to result from the breakdown of *protoxylem.

Carpel

A carpel is the flower's deepest component. It is typically surrounded by male reproductive organs known as stamens, which are also enclosed by petals. Flowers' petals are frequently big and highly colored. The petals are protected before the flower opens by sepals, which are usually green (Bourgaud et al., 2001).

One or more carpels can be found in a flower. If there are several carpels, they can be fused together or separated. A gynoecium, or pistil, is made up of all the carpels. A carpel consists of three structures:

- Stigma
- Style
- Ovary

Carpopodium

The carpopodium is the basal elongation of the achene wall that forms a hollow tubelike barren foot that may or may not be enlarged in relation to the fertile section.

Caruncle

The caruncle is a component located in Euphorbiaceae seeds' micropylar region. This structure promotes seed dissemination by ants (myrmecochory), although it is debatable if it also has agronomical significance by influencing seed germination.

Caruncle is a kind of elaiosome in the spurge family Euphorbiaceae. Seeds with a caruncle are caruncülate, while seeds without a caruncle are ecarunculate.

Caryopsis

A caryopsis (plural caryopses) is a form of simple dry fruit that is monocarpellate (produced from a single carpel) and indehiscent, similar to an achene except that the pericarp in a caryopsis is linked with the thin seed coat.

The caryopsis, also known as a grain, is a member of the Poaceae family, which also contains wheat, rice, and maize (Hadwiger, 2013).

Grain is frequently used as a term for cereal in a broader sense. Given that the fruit wall and seed are closely fused into a single unit, and the caryopsis or grain is a dry fruit, there is no need to distinguish the terms fruit and seed in these plant forms. Flower bracts are the "hulls" to be separated before processing in many grains.

Casual Alien

Invasion biology is a fast expanding branch of ecology. The discipline is densely packed with language used to explain various concepts and components. Unfortunately, some of these terms are misused, resulting in misunderstanding and confusion. The language employed in this discipline is primarily based on a study published in the journal Taxon in 2004 by Petr Pyek and others.

It should be noted that the classifications presented are not necessarily exclusive; for example, a species can be both foreign and naturalized.

Casual alien species can survive and even thrive for a short period of time, but they do not establish long-term populations. This means that, while these species' populations may survive and breed, they will eventually become extinct. To live, these species rely on recurrent imports into a region.

Cataphyll

In plant morphology, a cataphyll (also known as a cataphyllum or cataphyll leaf) is a reduced, small leaf. Many plants have both "real leaves" (euphylls), which do the majority of the photosynthesis, and cataphylls, which have been modified to fulfill a variety of specialized functions.

Cataphylls have bracts and bud scales, as well as little leaves that resemble scales. Cataphylls' roles, such as bud scales, may be restricted, and they are often shed once their job has been fulfilled (Guo & Liu, 2007).

In other cases, cataphylls serve a temporary function before dying and maybe shedding. Caducous organs shed early, a term that can apply to any organ that sheds early. Caducous stamens are seen in several Geraniums, for example. The sepals of Papaver species are shed at the same time as the petals appear, and they are an excellent example of caducous leaves.

Cataphylls come in a variety of shapes and sizes. Spines, corm-scales, and bud-scales, for example, may remain but not serve their primary function until death, whether physically shed or not. Live cataphylls include bulb-scales, rhizome-scales, cotyledons, and scaly bracts. Many of these groups come in a range of shapes and sizes. Bud-scales, for example, can be seen on numerous types of leaf or branch buds, as well as flower buds.

Catenulate

To be arranged in a series of rings or chains, as for spores

Catkin

A catkin, also known as an ament, is a slender, cylindrical flower cluster (a spike) with few or no petals that is wind-pollinated or sometimes insect-pollinated (as in Salix). They contain a high number of unisexual flowers that are tightly clustered along a central stalk that is usually drooping. They are found in several plant families, such as the Betulaceae, Fagaceae, Moraceae, and Salicaceae.

Catkins are produced by several trees and plants, including birch, willow, hickory, sweet chestnut, and sweetfern.

Many of these plants only have male flowers that form catkins, but the female flowers are single (hazel, oak), a cone (alder), or various shapes (mulberry). In other plants, both male and female flowers produce catkins (such as poplar).

They can be seen in the United Kingdom in January or February, when many trees are bare for the winter. They may even occur in December.

Figure 36: Detail of a male flowering catkin on a willow.

Source: https://en.wikipedia.org/wiki/Catkin

Caudate

- Having a narrow, tail-like appendage.
- Tailed at the apex.

Caudex

1: the stem of a palm or tree fern. 2: the woody base of a perennial plant.

A plant's caudex (plural: caudices) is a stem, although the phrase can also refer to a rootstock and, more specifically, a basal stem component from which new growth emerges.

In the strictest definition of the term, "caudex" refers to a stem, and is most commonly associated with plants that have a different stem morphology from the standard angiosperm dicotyledon stem: palms, ferns, and cycads are instances (Franklin et al., 2013).

Caudiciform, which simply translates "stem-like," is a synonym for pachycaul, which implies "thick-stemmed."

The term is derived from the Latin word caudex, meaning "tree trunk."

Caudiciform

Caudiciform plants have a caudex, which is a large, fat stem, trunk, or aboveground roots. Elephant's foot, Buddha belly, pregnant onion, and turtle back are just a few of the colorful names given to plants in this category. Caudiciform plants are numerous, and nearly as many gardening societies are dedicated to their odd architectural forms.

The diversity of form may amaze the inexperienced caudex lover. Caudiciforms can be found among the trees, vines, and even a few grasses and bulbs, in addition to the various bonsai and houseplant possibilities, many of which are succulents. Caudex selections can be observed in well-known genera like Euphorbia, Aloe, Cissus, Agapetes, and Cussonia, despite the fact that many caudiciform genera are scarce. The bulk of commercially developed caudiciforms come from the desert regions of Africa or Mexico. As a survival mechanism, the fat stems retain water when it is abundant for the plant to use all through long dry season. Many caudiciform plants are also classed as pachycauls, which is derived from the Greek word pachy, which means thick, and the Latin word caulis, which means stem. Pachycauls frequently have few branches.

Figure 37: Hottentot bread (Fockea edulis).

Source: https://www.pacifichorticulture.org/articles/caudiciform-plants/

Caulescent

In botany, having an obvious stem rising above the ground. Also cauliferous.

Cell

A cell is the most basic structure in all living systems, consisting of protoplasm surrounded by a cell membrane. All plants, including algae, and all animals are composed of cells that are organized in various ways to produce structure and function in an organism. Cells are classified into two categories by biologists: prokaryotic and eukaryotic. The structure of prokaryotic cells is simpler. They can only be found in single-celled organisms and some basic multicellular species. Eukaryotic cells are found in the majority of algae, as well as all higher plants, fungi, and animals (Eukarya domain). As a result, the discrepancies between these two cell types are critical to how an organism is classified, as well as an important issue in the evolutionary history of life on Earth (Foxcroft et al., 2017).

Almost all cells are so little that they cannot be seen with the human eye. There are obviously exceptions, but magnification is generally required to see cellular structures. In plants, a good hand-lens or loupe (as shown in the photo to the right) is sometimes sufficient, but a high power microscope is required when working with cells or analyzing how cells are clustered to form tissues and structures.

Cenanthous

Lacking both stamens and pistil, i.e. a flower with neither androecium nor gynoecium.

Ceraceous

Waxlike or waxy

Cernuous

Of a plant. : inclining or nodding: pendulous, drooping.

Cespitose

Caespitose plants are small, spherical, matted, and have the shape of a piece of turf.

(For instance, several or many stems in a tight tuft from a single rootstock, or several entangled rootstocks or roots jammed together in a cluster but not joined to each other.)

Chasmogamous

Open flowers with exposed anthers and stigma are known as chasmogamous flowers. When chasmogamous flowers reach maturity, they open. These blooms rely on pollinating agents to facilitate cross-pollination (and sometimes cross-pollination). Because the reproductive components of these flowers are frequently larger, more pollen must be produced. These flowers are showy and noticeable. Dichogamous blooms are common in this genus.

Cleistogamous flowers, on the other hand, are closed blooms that do not open and remain in the bud state. The anthers and stigmas are not revealed and lie close together in this kind. Because the reproductive structures are frequently smaller, less pollen is required (Fahad et al., 2017).

Because cleistogamous blooms never open, cross-pollination is impossible. As a result, in these blooms, only self-pollination is feasible. When the anther dehisces in the flower, the pollen grain comes into intimate contact with the stigma, causing self-pollination; consequently, no cross-pollination occurs and only self-pollination happens. These blooms are unobtrusive, inconspicuous, and monogamous.

In addition to chasmogamy and cleistogamy, several plant species have a mixed breeding strategy that employs both chasmogamous and cleistogamous blooms. This breeding system is also known as true cleistogamy, dimorphic cleistogamy, and the chasmogamous/cleistogamous mixed breeding system. In this mixed breeding system, species produce both chasmogamous and cleistogamous blooms on the same plant.

Chasmophyte

One of the most important types of vegetation grows on limestone rocks, which are home to a unique flora. Species that thrive in rock fissures must contend with an exceedingly hostile environment. As a result, in order to live, they have acquired many adaptations, such as strong roots. This keeps them afloat on the cliffs while also allowing them to make the most of the limited water and nutrients available in the soil (Elmer & White, 2018).

Furthermore, because there is no tall vegetation to cover it from high winds and direct sunshine, this ecosystem is vulnerable to these climatic variables. It is worth noting that the majority of chasmophytic species are indigenous.

Parnitha is a very steep mountain with many cracks and gorges, particularly on its south side. This encourages the formation of various chasmophytic communities with endemic species, such as the Parnitha bellflower - *Campanula celsii ssp. parnesia*, and the Parnitha wild carnation (endemic to Parnitha)

Chimera

Chimera, often called Chimaera, is a plant or plant part that is a hybrid of two or more genetically distinct cell types.

A chimera could be a "graft hybrid," which is a bud that forms at the intersection of the scion and stock in plant grafting and contains tissues from both plants. Although similar chimeras emerged inadvertently in the past, the German botanist Hans Winkler was the first to seriously study them in 1907. In his first experiments, he grafted black nightshade (*Solanum nigrum*) on tomato (*Solanum lycopersicum*), and at the nexus, all the shoots were either nightshade or tomato except one, which arose at the junction of the two tissues and had the characteristics of both nightshade and tomato. Because it had part of one species and part of another, Winkler dubbed this shoot a chimera (De Waele & Elsen, 2007).

In subsequent trials, he gave specific names to some of his graft hybrids. Another botanist, Erwin Baur, later supplied proof that two plants to which Winkler had given unique names were constructed of a tomato core with a nightshade skin one and two cell layers thick, respectively, and two others of a nightshade core with tomato skins one and two cell layers thick. Thus, the components of a chimera retain their identity while being arranged in a precise manner at the growth point.

Chimeras can also form as a result of a mutation in cells in an expanding region. The new tissue may be visibly different from the old (for example, it may be colorless instead of green), but far more often, the difference is only apparent upon careful examination, such as when the number of chromosomes is altered.

Chiropterophilous

Nectar- and pollen-eating bats pollinate over 500 tropical types of plants, and they have acquired specific characteristics to make their nectar and pollen enticing to nocturnal fliers. These plants are chiropterophilous, which means "bat-loving" (bats being mammals of the order Chiroptera). Plants that rely heavily on bat pollinators cater to them by producing massive, white flowers that bats can easily identify at night. The flowers have a sour or musky odor and open after midnight, when bats emerge from their day roosts to feast. Many bat-pollinated flowers are shaped like vases to fit a bat's face, whereas others are flattened and brushy to load pollen onto a bat.

Chiropterophilous plants can even create compounds that are useful to the bat but not to the plant. Because bats ingest both the pollen and nectar of their blooms, the pollen of bat-loving plants is high in protein and contains the amino acids tyrosine and proline, which are vital for bat health. Proline is essential for the construction of strong wing and tail membranes, whereas tyrosine is required for the production of milk (Chen & Yamaguchi, 2002).

Nectar-eating bats have their own set of adaptations. They have fleshy bristles on their long tongues, like many bees, to scoop out pollen and nectar. They have sharp vision and a good sense of smell, but their sonar is sometimes damaged. Migratory bats pollinate a wide variety of species as they move, and plants have been observed to flower in sequence along a form of "nectar corridor" that corresponds to the bats' migratory route.

Chlorophyll

Chlorophyll (also chlorophyl) is a green pigment found in cyanobacterial mesosomes in algae and plant chloroplasts. Photosynthesis, which allows plants to absorb light energy, requires chlorophyll.

Figure 38: Chlorophyll is responsible for the green color of many plants and algae.

Source: https://en.wikipedia.org/wiki/Chlorophyll

Chlorophylls absorb the most light in the blue and red ranges of the electromagnetic spectrum. It, on the other hand, absorbs very little of the near-green portions of the spectrum. Green light is less assimilated when diffusively reflected by components such as cell walls, hence chlorophyll-containing tissues appear green. There are two types of chlorophyll in green plant photosystems: chlorophyll a and chlorophyll b. The chlorophyll reaction center receives light energy and transfers it to other photosystem components (Clemens et al., 2013). In a process known as charge separation, the absorbed energy of a photon is transformed to an electron. When an electron is taken from chlorophyll, one oxidation step happens. The high-energy electron is donated by chlorophyll to an electron transport chain, which is a series of chemical intermediates. By absorbing an electron removed from water, the charged chlorophyll reaction center (P680+) is then reduced to its ground state. The electron utilized to reduce P680+ is generated from several steps of water oxidation into O2 and H+.

This reaction produces O2 gas in photosynthetic organisms such as plants, and it accounts for approximately all of the O2 in the Earth's atmosphere. Because Photosystem I frequently works in series with Photosystem II, its P700+ is normally lowered when it accepts electrons from Photosystem II via many precursors in the thylakoid membrane. Electron transfer reactions in thylakoid membranes, on the other hand, are complex, and the source of electrons employed to decrease P700+ can vary.

Chloroplast

Chloroplasts are photosynthesis organelles in plant and algal cells where the photosynthetic pigment chlorophyll absorbs sunlight energy, converts it, and stores it in the energy compounds ATP and NADPH while freeing oxygen from water. The ATP and NADPH are subsequently used in the Calvin cycle to produce organic molecules from carbon dioxide. Chloroplasts also play a number of other roles in plants, such as fatty acid synthesis, amino acid synthesis, and immune response. The number of chloroplasts per cell in unicellular algae ranges from one to 100 in plants such as Arabidopsis and wheat.

A chloroplast is a kind of plastid organelle defined by two membranes and a high chlorophyll content. Other types of plastids, such as leucoplast and chromoplast, have less chlorophyll and do not undergo photosynthesis.

Chloroplasts are highly active, circulating and migrating throughout plant cells and occasionally pinching in two to reproduce. Light hue and intensity, for example, have a significant impact on their behavior. Like mitochondria, chloroplasts have their own DNA, which is thought to have been carried down from their ancestral species photosynthetic cyanobacterium ingested by an early eukaryotic cell. Chloroplasts cannot be created by the plant cell and must be transferred down by cell division to each daughter cell (Craigie, 2011).

With the exception of the amoeboid *Paulinella chromatophora*, all chloroplasts are most likely the result of a single endosymbiotic event in which a cyanobacterium was engulfed by a eukaryote. Despite this, chloroplasts can be found in a very diverse range of species, some of which are not even directly linked to one another—the result of several secondary and even tertiary endosymbiotic occurrences.

The word chloroplast is derived from the Greek words chloros, which means green, and plastes, which means "the one who forms"

Chlorosis

Chlorosis is distinguished by the yellowing of leaf tissue produced by a lack of chlorophyll. Chlorosis can be caused by a variety of factors, including waterlogging, injured roots, compacted roots, elevated alkalinity, and nutrient deficiencies in the plant. Nutrient shortages can arise due to a lack of nutrients in the soil or because the nutrients are inaccessible due to a high pH. (Alkaline soil). Alternatively, nutrients may not be absorbed as a result of wounded roots or insufficient root growth.

One of the more prevalent nutrients related to chlorosis is a deficiency of iron. Chlorosis can also be caused by a lack of manganese or zinc in the plant. To distinguish an iron shortage from a zinc or manganese deficiency, look at which leaf went chlorotic first. Iron chlorosis begins on the younger or terminal leaves and progresses to the older leaves (Chandler & Sanchez, 2012).

Manganese and zinc deficiencies, on the other hand, first emerge on the inner or older leaves and then move outward. Iron is essential for plant chlorophyll formation. Chlorophyll is responsible for the green color of leaves and is essential for the plant to produce the food it needs to grow. Iron is also needed for a number of enzyme functions that control plant metabolism and respiration. Iron becomes less soluble as the pH of the soil goes above 6.5 to 6.7. (7.0 is neutral - below 7.0, the pH is acidic; above 7.0, the pH is alkaline). Only when the pH is between 5.0 and 6.5 can most plants take iron as a free ion (Fe++).

Other elements in high concentrations in the soil, such as calcium, zinc, manganese, phosphorus, or copper, can bind to iron, rendering it unavailable to the plant.

However, a lack of potassium in the plant reduces the plant's availability of iron. Inadequate iron levels in the soil are also an issue. Most soils in the Chicagoland area have appropriate iron levels. The issue is that the plant does not have access to the iron in the soil. The majority of soils in Northeastern Illinois were created from limestone bedrock. As a result, chlorosis is frequently caused by high soil pH.

Chlorosis is a disease that affects both herbaceous and woody plants.

Cirrhose

(Botany, of leaves) Ending in a tendril at the apex. (A thin, spirally coiling stem that attaches a plant to its support.

Cladode

Cladodes (also known as cladophylls or phylloclades) are shoot systems in which no leaves develop; instead, the stems flatten and take on the plant's photosynthetic duties. True leaves are the scales found on asparagus stalks (*Asparagus officinalis*; Asparagaceae). If the thick, meaty asparagus spears continue to grow, flat, green, leaflike structures known as cladodes form in the scale leaf axils. The existence of cladodes in disparate desert angiosperm families is a prime example of convergent evolution, or the independent emergence of the same trait in unrelated taxa (Zehnder et al., 2007).

Many desert members of the spurge (Euphorbiaceae) and milkweed (Apocynaceae) families exhibit comparable vegetative morphologies that are acquired through changing distinct components to look and function in the same way. Columnar, water-storing green stems, reduced leaves, and defensive spines or thorns characterize each of these plant families. They are also known as stem succulents. The leaves on the main stems of cacti last just a short period (they do not even mature as scale leaves), and the leaves of axillary buds (the spherical cushion portions, or areoles, on the trunks) grow as spines. The leaves on the main stems of many Euphorbiaceae members are green but short-lived, and the stipules develop as spines. The leaves of many Apocynaceae plants are also tiny and transitory, and the axillary buds grow into thorns.

Cacti are virtually entirely New World plants suited to dry or desert conditions, although the Euphorbiaceae and Apocynaceae are found in Asia and Africa in analogous environments. Leaf reduction is so extreme in the Cactaceae that epiphytic cacti (e.g., Epiphyllum) of the Neotropics can no longer produce leaves; alternatively, they create thin, flat cladodes that superficially mimic leaves.

Clavuncula

The clavuncula is an expanded stigma in Apocynaceae, with the sides and bottom surface serving as the receptive zone; it is usually continuous with anthers.

Apocynaceae is an angiosperms family that comprises trees, shrubs, herbs, stem succulents, and vines (from Apocynum, Greek for "dog-away"). Because numerous taxa are used as dog poison, it is frequently referred to as the dogbane family (Yamada et al., 2003).

Members of the family are endemic to Europe, Asia, Africa, Australia, and North America, with a few temperate members.

Cleistogamous.

Cleistogamy is a sort of automatic self-pollination that occurs in certain plants that can reproduce by employing non-opening, self-pollinating flowers. This activity is most common in the grass family and is particularly well known in peanuts, peas, and pansy. Viola, on the other hand, is the biggest genus of cleistogamous plants.

Chasmogamy, or "open marriage," is the more prevalent reverse of cleistogamy, or "closed marriage." Almost every plant that produces cleistogamous blooms also produces chasmogamous flowers. The main advantage of cleistogamy over chasmogamy is that it uses less plant energy to generate seeds because the growth of petals, nectar, and enormous volumes of pollen is not needed. Because of its efficiency, cleistogamy is particularly effective for seed generation in poor locations or under bad conditions. After being badly injured by grazing, Impatiens capensis, for example, has been seen to produce only cleistogamous blooms and to establish populations on unsuitable sites with only cleistogamous flowers (Yakhin et al., 2017).

The obvious downside of cleistogamy is that it causes self-fertilization, which may prevent the development of genetically superior plants.

Researchers are seeking to use cleistogamy to limit gene flow in genetically modified (GM) rapeseed in order to reduce mixing of GM and non-GM crops. However, initial evauations from Co-Extra, a current EU research program, reveal that, while cleistogamy decreases gene flow, it is not yet a consistently dependable technique for biocontainment; due to the cleistogamous trait's instability, certain blooms may open and distribute GMO pollen.

Climber

Climbers are significantly more evolved than creepers. Climbers have a thin, long, and weak stem that cannot stand alone, yet they can grow vertically and support their weight by using external support. Climbing tendrils are specialized structures that these plants use. Other climbing plant names include pea plant, grapevine, sweet gourd, money plant, jasmine, runner beans, green peas, and so on (Yakhin et al., 2017).

Gardeners can take advantage of climbing plants' proclivity to develop swiftly. A climber can quickly create a plant display if one is desired. Climbers can be manipulated to climb over walls, fences, and other structures. Climbers can be grown on top of other plants to provide interest. Artificial assistance can also be offered. Some climbers can do it on their own; others require assistance, such as tying them in and instructing them.

Vines vary greatly in size, shape, and evolutionary origin. Charles Darwin classified climbing groups according on how they climbed. He classified five classes of vines – twining plants, leaf climbers, tendril bearers, root climbers and hook climbers.

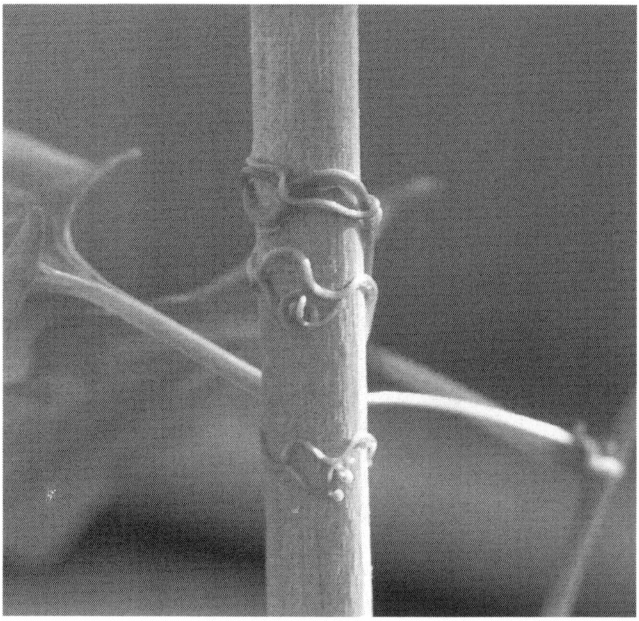

Figure 39: Tendril-supported vine (*Brunnichia ovata*).

Source: https://en.wikipedia.org/wiki/Vine#/media/File:Brunnichia_ovata.jpg

Clone Plant

Plant cutting, also known as cloning, is a method of vegetatively (asexually) reproducing plants that entails placing a section of the donor plant's stem or root in moist soil, potting mix, coir, or rock wool. The cutting grows new roots, stems, or both and therefore becomes a separate plant from the parent.

Many plant vegetative components can be employed. The most popular ways are as follows:

- Stem cuts, in which a portion of the stem, including at a minimum of one leaf node, is buried in the soil. The incision might form new roots, ordinarily around the node.
- Root cuttings, where a piece of the root is buried just beneath the soil surface and creates new shoots (Wisler & Norris, 2005).
- Scion cuttings, often known as dormant ligneous woody twigs.
- Eye cuttings are foliated or defoliated stalk portions with one or more eyes.

- Leaf cuttings are formed by placing a leaf on moist soil. These must grow new stems as well as new roots. Some leaves will develop a single plant at the leaf's base. In some cases, multiple new plants can be formed at multiple locations on a single leaf In some species, and these can be caused by cutting the leaf veins.

Although certain species, such as willow, blackberry, and pelargoniums, can be grown simply by inserting a cutting into damp soil, the majority of species require more attention. Most species need humid, warm, partially shady conditions to strike, mandating the above-mentioned method. Particularly difficult plants may necessitate cool air above and warm soil below. In addition, for many more challenging cuttings, use the sort of cutting that has the best chance of survival with that specific plant species.

Figure 40: Plant cuttings can be taken by cutting a stem of a plant, or a single leaf.

Source: https://en.wikiversity.org/wiki/Plant_propagation/Cloning

Coalescent

The union of plant parts of the same kind such as the united sepals of flowering plants.

Coenobium

- The joining of similar plant parts, such as the united sepals of flowering plants.
- a usually spherical colony of unicellular organisms surrounded by a common investment
- especially: a colony having a definite number and specific arrangement of cells.

In contrast to a syncytium, which is generated by cellular aggregation accompanied by the breakdown of the cell membranes within the mass, a coenocyte is a multinucleate cell that can be produced by numerous nuclear divisions without cytokinesis. The term syncytium alludes to the coenocytic blastoderm of invertebrates in animal embryology. A coenobium (plural coenobia) is a coenocytic cell, and most coenobia are made up of a specified number of cells, typically a factor of two

According to studies, coenobium production may be a grazing resistance in some plants.

Coleoptile

Coleoptile is the sharp defensive sheath that surrounds the forming shoot in monocotyledons like grasses, encompassing only a few leaf primordia and the monocot embryo's shoot apex. The coleoptile protects the first leaf as well as the growing stem in seedlings, allowing the first leaf to emerge. Coleoptiles have two vascular bundles on each side of their bodies. Unlike the rolled-up flag leaves within, the pre-emergent coleoptile does not produce much protochlorophyll or carotenoids, so it is generally fairly pale. Some preemergent coleoptiles, on the other hand, collect purple anthocyanin pigments (Weitemier et al., 2014).

Coleoptiles are composed of cells that have all been specialized for rapid stretch growth. They do not split, but rather grow larger as more water is absorbed. Coleoptiles also include water vessels (typically two) that run along the axis and provide a source of water.

When a coleoptile rises to the top, it stops developing and the flag leaves puncture its top, allowing it to continue developing. Wheat coleoptile development is greatest on the third day of germination (if in the dark).

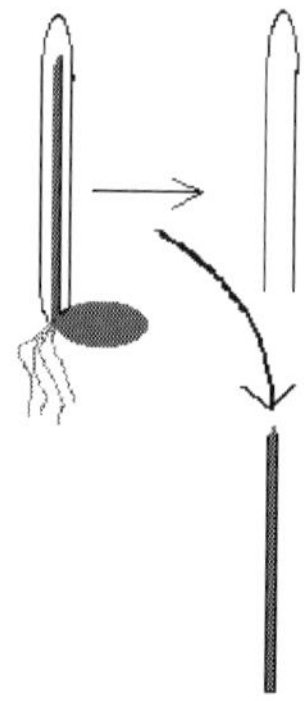

Figure 41: Schematic image of wheat coleoptile (above) and flag leaf (below).

Source: https://en.wikipedia.org/wiki/Coleoptile

Coleorhiza

The coleorhiza, coleorriza, or root sheath is a surface of tissue that covers the root in the seed and discourages root rot in monocotyledons. It is a protective covering that wraps around the root. The coleorhiza is the first component to emerge from the seed. During germination, the coleorhiza first expands through cell elongation, but is finally pierced through the root and persists as a collar around the root base. A coleorhiza also exists in the adventitious roots.

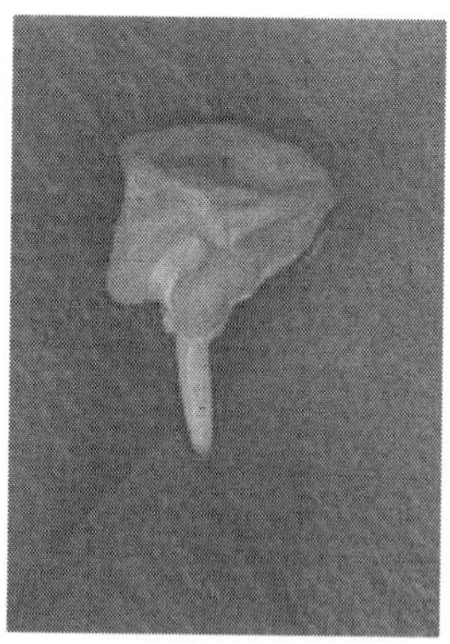

Figure 42: Coleorhiza with sweet corn.

Source: https://en.wikipedia.org/wiki/Coleorhiza

Collenchyma

Collenchyma is a plant support tissue composed of living elongated cells with irregular cell walls. Collenchyma cells have a lot of cellulose in their cell walls, which makes them look polygonal in cross - section. The tissue's strength is enhanced by stronger cell walls and longitudinal interconnecting of cells. Collenchyma is one of three ground, or fundamental, tissues of plants, along with parenchyma and sclerenchyma (Wilkinson et al., 2012).

The fact that collenchyma is exceedingly plastic—the cells can stretch and so respond to greater organ growth—is a significant traitThe tissue is typically found in the cortex of foliage, and it is the primary supporting tissue in many herbaceous plants. In secondary-growing plants, collenchyma tissue is only viable for a short time before becoming squashed when woody tissue develops. It is typically seen in the ridges and angles of stems, as well as the veins of eudicot leaves. Celery stalk "strings" are a good example of collenchyma tissue.

Colleter

Colleters are plant structures, multicellular secretory hairs, found in groups near the base of petioles, on stipules, and on sepals. They are found in members of the Loganiaceae and Rubiaceae families.

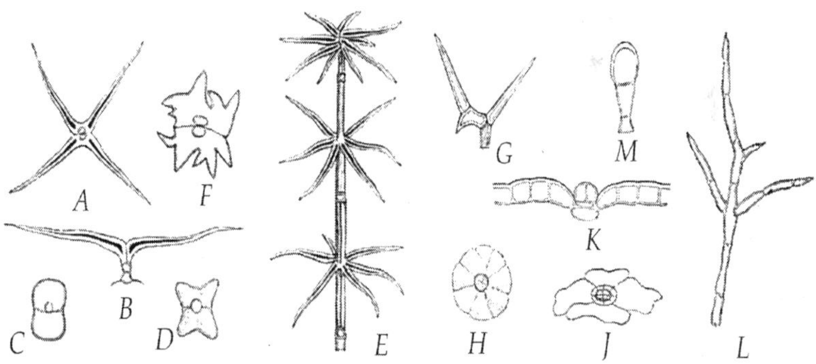

Figure 43: Several hair types occur on plants of Loganiaceae, including multicellular secretory colleters shown here in cross-section (C, D, H).

Source: https://en.wikipedia.org/wiki/Colleter_(botany)#/media/File:Hairs_of_Loganiaceae_EP-IV2-012.png

Columella

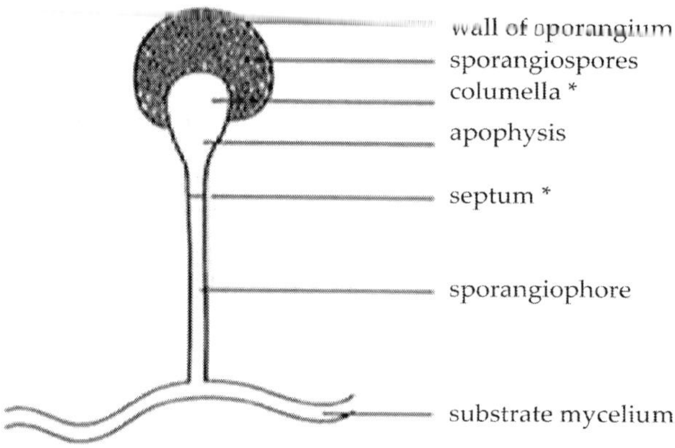

Figure 44: Zygote fungus sporangium, with columella labeled.

Source: https://en.wikipedia.org/wiki/Columella_(botany)#/media/File:Sporangium..png

Columella (in plants) is an axis of sterile tissue that runs through the heart of a moss' spore-case. It is a centrally vacuolated portion of a hypha that bears spores in fungus. Myxomycetes use the term in a similar way.

The term columella is also used to allude to the cells in the root cap from S1 – S4 that are located apically of the quiescent center.

Cone

A cone, also known as a strobilus in botany, is a clump of scales or bracts that is generally ovate in shape and houses the sexual organs of particular nonflowering plants. All gymnosperms, some club mosses, and horsetails have the cone that separates them from other conifers.

A conifer cone (formally strobilus, plural strobili) is a reproductive organ found on plants in the Pinophyta (conifers) division. The well-known woody cone is the female cone, which produces seeds. Even when fully mature, the male cone, which releases pollen, is normally herbaceous and much less noticeable. The term "cone" is derived from the Greek word konos (pinecone), which also inspired the geometric cone. The various pieces that make up a cone are referred to as scales. A conifer cone's umbo refers to the first year's development of a seed scale on the cone, which appears as a protuberance at the end of the two-year-old scale (Waraich et al., 2011).

A cone is a reproductive structure found in gymnosperms that is properly known as a strobilus. It is made up of sporophylls that contain sporangia that produce spores. Male and female cones are produced differently by gymnosperms. The huge woody female cones of pines, firs, and other conifers are composed of structures known as ovuliferous scales, which contain the ovules. Clubmosses and horsetails also generate cones.

Corcle

The heart of the seed; the embryo or germ.

Cordate

Heart -shaped, such as the leaves of the sweet violet (Viola odorata) or the lemmas of quaking grasses

The Eastern Redbud is a classic example of cordate.

Figure 45: Eastern Redbud.

(Cercis canadensis)

Source: http://woodyplantstutorial.nres.illinois.edu/shape/sha-cordate.html#

From the subopposite leaf arrangement to the crenate margins and the cordate leaf shape, the Katsuratree is a useful plant for identification features. Here the shape is cordate, but the leaves vary and some can appear orbicular.

Corolla

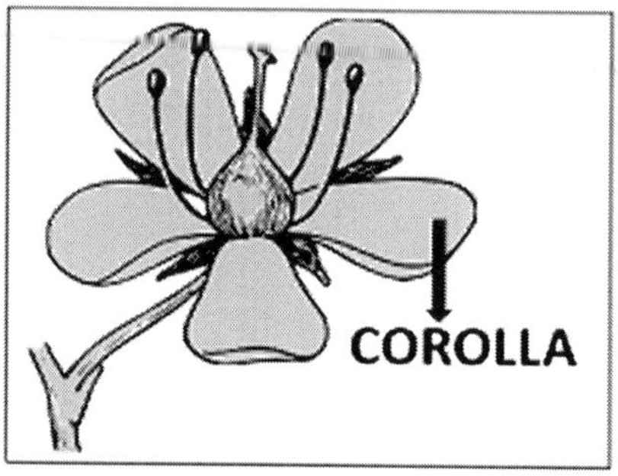

Figure 46: Corolla.

Source: https://biologyreader.com/corolla-in-plants.html

In plants, a corolla is a bunch of colored petals that encircles the reproductive components of a flower, such as stamen and carpel. As a result, corolla or petals simply refer to the second whorl of a flower, which is located inside the calyx.

In plants, the corolla is a second whorl that, along with a layer of the calyx, completes the perianth. It often consists of many petals, the principal function of which is comparable to that of the calyx, which shields the flower's reproductive components (Van Dijk et al., 2021).

The corolla is a sterile portion of the flower as well. It might be tubular, funnel-shaped, wheel-shaped, or any other shape. The corolla is larger in size than the calyx. Petals are brightly colored and fragrant, and their main purpose is to attract pollinators like as bees and birds.

Corona

A corona is a crown-shaped, funnel-shaped, or trumpet-shaped outgrowth or appendage of a flower's perianth, such as that of a daffodil. Also known as a crown.

Cotyledon

A cotyledon is a crucial component of a plant's embryo within the seed, and it is defined as "the embryonic leaf in seed-bearing plants, one or more of which develops first from a germinating seed." The number of cotyledons present is one of the criteria used by botanists to identify blossoming plants (angiosperms). The term "monocotyledonous" refers to species that have only one cotyledon ("monocots"). Two embryonic leaves are present in dicotyledonous plants ("dicots") (Van Bruggen et al., 2016).

Cotyledons behave similarly to leaves in dicot seedlings with photosynthetic cotyledons. True leaves and cotyledons, on the other hand, grow independently. Cotyledons, as well as root and shoot meristems, are formed during embryogenesis and hence exist in the seed prior to germination. True leaves, on the other hand, are formed post-embryonically from the shoot apical meristem, which is in charge of the plant's succeeding aerial parts.

Figure 47: Sprouting of a monocot versus a dicot. The visible section of the monocot plant (left) is the meristem's first genuine leaf; the cotyledon itself stays within the seed.

Source: https://en.wikipedia.org/wiki/Cotyledon

The cotyledon is a highly modified leaf that consists of a scutellum and a coleoptile that is present in grasses and several other monocotyledons. The scutellum is a seed element that collects stored nutrients from the adjacent endosperm. The coleoptile is a cap that rests on top of the plumule to shield it (precursor to the stem and leaves of the plant).

Craspedodromous

A type of pinnate venation where the secondary veins terminate at the leaf margins.

Figure 48: The secondary veins branch and are termed semicraspedodromous. If the secondary veins are unbranched and terminate at the leaf margin it is simple craspedodromous.

Source: http://plant-phytography.blogspot.com/2011/12/craspedodromous-type-of-pinnate.html#:~:text=craspedodromous%20krasp'e%2Ddo%2Ddro'mos%20adj.&text=A%20type%20of%20pinnate%20venation%20where%20the,terminate%20at%20the%20leaf%20margins.&text=The%20secondary%20veins%20branch%20and,leaf%20margin%20it%20is%20simplecraspedodromous.

Cremnophyte

Any plant that is adapted to growing on cliffs, especially in crevices.

Crisped

Finely curled, as with the edges of leaves and petals.

Cryptogam

A cryptogam is a plant or plant-like organism that propagates by spores rather than flowers or seeds. The term Cryptogamae means "hidden reproduction," and refers to the fact that no seed is generated; thus, cryptogams are plants that do not produce seeds. Other names used on occasion include "thallophytes," "lower plants," and "spore plants." The seed plants, Cryptogamae, are the polar opposite of Phanerogamae or Spermatophyta (Arruda et al., 2015). The most well-known cryptogams are algae, lichens, mosses, and ferns, but it also includes non-photosynthetic organisms that are generally categorized as plants, such as fungus, slime molds, and bacteria. This classification is no longer used in Linnaean taxonomy.

Culm

Figure 49: Culms of sugarcane.

Source: https://en.wikipedia.org/wiki/Culm_(botany)

A culm is a grass or sedge's aerial (above-ground) stem. It is derived from the Latin word culmus, which means "stalk," and it originally referred to the stem of any sort of plant.

Cultigen

A cultigen is a plant that has been artificially selected or purposely transformed by humans. Most of these plants have an economic value in gardening, agriculture, or forestry. Plants that fit these criteria remain cultigens even if they are naturalized in the wild, purposefully planted in the wild, or grown in culture, because cultigens are defined by their mode of origin instead of where they grow.

Cultivar

A term derived from "cultivated variety" that refers to a collection of cultivated plants that can be distinguished by one or more traits (morphological, physiological, cytological, chemical, or other). When replicated, the assemblage retains its own traits (either sexually or asexually). Cultivars can develop in the laboratory or be imported from the wild. It is a horticulturally interesting or valuable variety. Cultivar names are surrounded by single quote marks, such as 'Blue Carpet' or 'Alba.' All new names created after 1959 must be in common language, however, names created in Latin before this date are kept in Latin form (Bourgaud et al., 2001).

Cuticle

A plant cuticle is a protective coating that covers the epidermis of leaves, new shoots, and other periderm-free aerial plant organs. It is made up of lipid and hydrocarbon polymers impregnated with wax and is produced solely by epidermal cells.

Cutting

In horticulture, a plant cutting is a part of a plant used for vegetative (asexual) propagation. In a suitable medium, such as wet soil, a bit of the donor plant's stem or root is put. If the conditions are favorable, the plant fragment will begin to grow independently of the parent plant, a process known as striking. A stem cutting produces new roots, whereas a root cutting produces new stems. Some plants may grow from leaf fragments, producing both branches and roots. These are referred to as leaf cuts. Scions, which are used in grafting, are also known as cuttings.

Cyathium

A cyathium is a specialized pseudanthia ("false flowers") that form the inflorescence of Euphorbiaceae. A cyathium is made up of:

Five bracteoles (rarely four). These are tiny, joined bracts that create a cup-shaped involucre. Their top tips are unattached and initially cover the aperture of the involucre (like the shutter of a camera). These are alternated with:

Five (1–10) nectar glands, which are occasionally merged.

One severely reduced female flower stands at the base of the involucre, consisting of an ovary on a short stem with pistil, and is surrounded by:

Five groups of extremely reduced male flowers, each consisting of a single anther on a stem.

Cyathophyll

In Euphorbia, the bract-like structure on which the involucre sits, usually but not always occurring in twos. They may sometimes be brightly colored and confused with petals.

Cypsela

A type of dry, one-seeded, indehiscent fruit formed from an inferior ovary.

D

Deciduous

Figure 50: Red deciduous tree shedding its leaves.

Source: https://study.com/academy/lesson/deciduous-trees-lesson-for-kids-definition-facts.html

Deciduous trees drop their leaves in fall to prepare for the successive winter. Their leaves have cells that resemble scissors, slicing them off from the primary piece of the tree during autumn. The incision is then shut. Thus, the winter freeze is fixed out, while valuable water is fixed in

(Van Breusegem et al., 2001). The vivid leaves look marvelous whereas their trees are as yet wearing them, however, they can cause issues when they drop onto the grass. Based on the plant variety you may need to tidy up different plant types, varying from pods to pine needles. Less messy deciduous trees include the Sunburst honey insect, commonly known as *Gleditsia triacanthos* var. inermis Suncole).

Decorticate

Decortication is the process of dividing the plant bast from the internal core. Bast filaments acquired from dicotyledonous plants like jute, hemp, flax, ramie or kenaf, have various commercial applications. Bast strands have generally been used by the material business to fabricate products like cover yarn, rope, geotextile, traditional hessian and burlap. Be that as it may, increasingly bast filaments are being utilized by different ventures for such things as the production of non-woven mats and covers, composite materials, and car interior boards. Cellulose-based enterprises are likewise recovering cellulose from bast for use in mash and paper creation. Before bast filaments can be utilized, they should be first isolated from the internal center of the fiber plant through the decortication process.

Decumbent

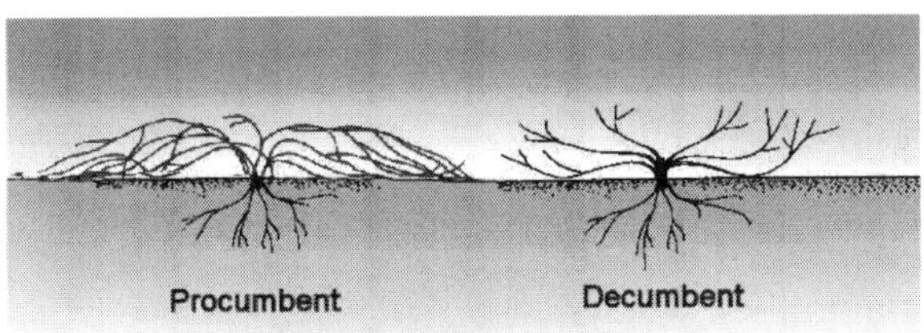

Figure 51: Decumbent plants have upward facing shoots.

Source: https://www.cactus-art.biz/note-book/Dictionary/Dictionary_D/dictionary_decumbent.htm

Decumbent is a type of blossoming plant in the Melanthiaceae family. A woody plant consisting of branches spreading across the ground or lying level on the ground, or stretching across the ground for the greater part of its length yet with apex turned upwards.

Decumbent is a tiny, round, apple green leaf that remains firm against adaptable red stems. This simple to grow cultivar branches openly and spills pleasantly from pots. In ice-free environments, it can likewise be grown outside in holistic to the partial sun as a diffusion ground cover. Portulacaria will more often than not be hesitant to sprout in development, however, they can create bunches of tiny purple roses during midsummer (Chandler & Sanchez, 2012).

Delicate succulents won't endure hard ice, yet assuming that there is a danger of frigid temperatures they can be brought inside to mature on a radiant window ledge, or below a grow light. They require sufficient daylight, sufficient drainage, and sporadic water to forestall rot. Pick containers comprising of drainage openings to utilize quick-draining cactus and porous soil with half to 70% mineral coarseness like rough sand, pumice, or perlite.

Decurrent

Decurrent is a terminology utilized in plant science and mycology to define plant or fungal parts that expand downward. The "decurrent branching propensity" is a plant structure normal for bushes and various angiosperm trees, compared to the excurrent or "cone-formed crown" typical among gymnosperms. It additionally relates to plant parts that stretch downward, frequently applied to leaf edges that partially fold or have wings over the stem or petiole and stretch out down along the stem. Having the leaf base stretching out down the stem beneath the inclusion. Decurrent growth is brought about by weak apical predominance. Apical predominance basically implies that the vitally focal stem grows quicker than all the other parts. Decurrent leaves have leaf edges that fold themselves over a length of the stem.

Dehiscent

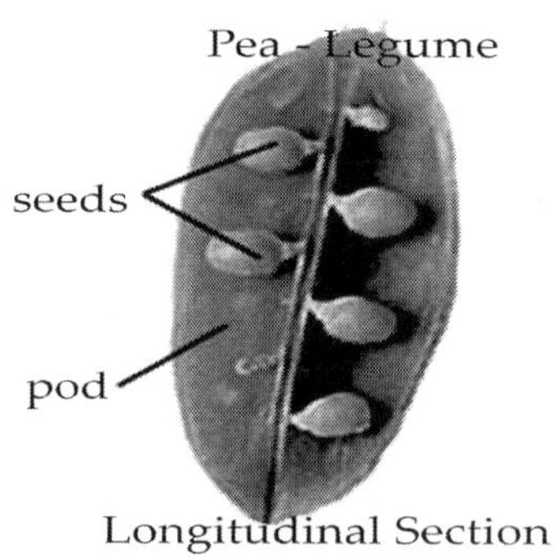

Figure 52: *Dehiscent fruits have a weak line that pops and disperses the fruits inside.*

Source: https://science.jburroughs.org/resources/flower/fruit2.html

Fruits are dehiscent when the pericarp parts open at development and deliver the seeds, or indehiscent in case the pericarp stays unblemished when the fruit is released from the plant. The three principal forms of dehiscent fruits include follicles, capsules and legumes. Dehiscent means plants that open upon development. These plants have a specific line of weakness that split open to deliver the internal content. Dehiscence is typical among sporangia, fruit and anthers. Sometimes, the structure may totally extract from the plant. Taking everything into account, the external covering needs to break to permit the seeds to scatter and grow (Van der Kooi & Ollerton, 2020).

Dendroid

Dendrology refers to the study of trees, and the people who do the research are called dendrologists. Thus, dendroid defines something that "branches" every way from a focal "trunk" irregularly. The term is quite often utilized by biologists, who regularly talk about dendroid seaweeds, dendroid green growth, and dendroid moss.

Deserticolous

Deserticolous plants thrive in the desert. Desert plants will generally appear to be unique from plants local to other areas. They are frequently enlarged, spiked, and have minuscule leaves that are seldom bright green. Their unusual appearance is an aftereffect of their amazing transformations to the difficulties of the desert environment. Aridity is the sole variable that characterizes a desert and is the essential constraint to which desert life forms should adjust. Desert plants have created three fundamental adaptability techniques: succulence, drought aversion, and resilience. Every one of these is an alternative but compelling set-up of transformations for succeeding under conditions that would otherwise kill plants from different locations.

The drought-tolerant shrub and tree roots are broad contrasted with those of plants in wetter environments, encompassing a region twice the distance across the canopy. They exploit soils at a more prominent depth than the succulent roots; some of the time they stretch out to outrageous depths (e.g., mesquite). The vast majority of a mesquite's roots, be that as it may, are around three feet (0.9 m) of the surface. Succulents can retain water just when the soil is almost immersed. Interestingly, drought-resistant plants can retain water from soil that is a lot drier. Additionally, these plants can photosynthesize with reduced leaf moisture substances that would demonstrate lethality to most plants.

Diadelphous

The stamens of specific blossoms, like those of certain individuals from the bean family, are diadelphous. (Plant science) Having its stamens intertwined partially by the fibers so they form two separate packs, or a group and a single separate stamen. In 'diadelphous' condition, the fibers of nine stamens are joined into one pack and the tenth back stamen stands separated. This is the trademark feature of the family 'Papilionaceae'. For instance, pea (*Pisum sativum*).

Diaspore

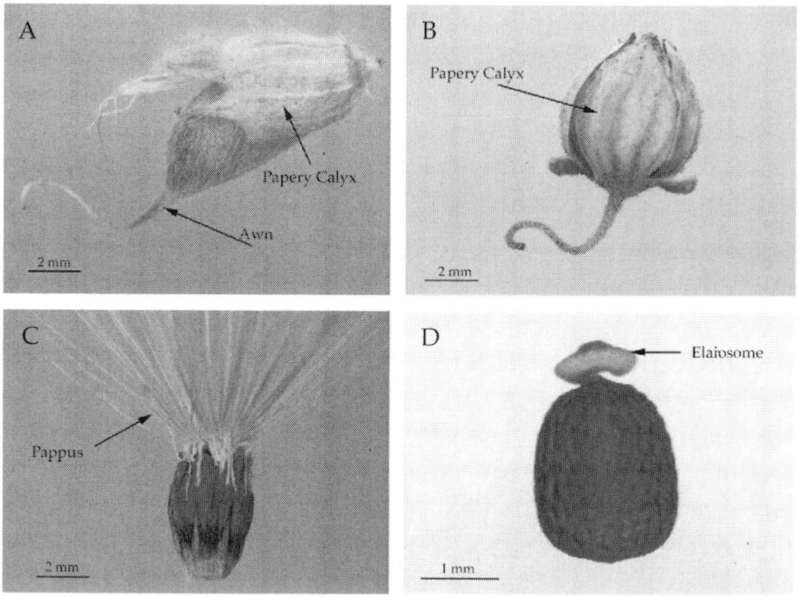

Figure 53: (A) pod of *Astragalus granatensis* and other Diaspore types.

Source: https://www.researchgate.net/figure/Diaspore-types-A-pod-of-Astragalus-granatensis-B-capsule-of-Helianthemum-croceum_fig9_229366887

In natural science, a diaspore is a plant dispersal section comprising of a spore or seed, in addition to any extra tissues that help with dispersal. For some seed plants, diaspore is both a seed and fruit product, or a seed and elaiosome. In a couple of seed plants, the diaspore is most or the entirety of the plant, and is known as a tumbleweed. While a diaspore (or 'disseminule') can be described as 'a reproductive plant part, like a seed, organic product, or spore that is changed for dispersal', the definition is generally expanded to incorporate any plant part that could bring about the foundation of another plant. Subsequently, it incorporates bryophyte spores, yet in addition sections of established plants, as well.

Inspecting the plumage of bird species in their Arctic nesting places – preceding their South Pole-ward movement – researchers tracked down cases of diaspores of bryophytes, yet in addition to green growth/cyanobacteria, and parasites. The existence of these putative propagules among bird plumages consequently appears to set up this peculiarity as one more case of diaspore (delivery of plant – and green growth/fungi/microbes– dispersal units on the outer surface of an organism) (Twidwell et al., 2013).

However, in light of the fact that these travelers might be available toward the beginning of the travel doesn't really imply that they show up at the transporter's destination, like the semipalmated sandpiper and red phalarope – is the southernmost point of South America; for example could the disseminules be devoured during trimming as a kind of in-flight nibble by the birds.

Dichasium

A dichasium is one part of a cyme and is defined by an obstructed focal blossom and two sidelong blossoms on extended pedicels, like the wood stichwort (*Stellaria nemorum*). Different definitions include: the cyme having two parallel blossoms or branches starting from inverse points underneath a terminal bloom. A cyme where two inverse branches emerge underneath every terminal blossom, and a cyme with two parallel blossoms or branches beginning from inverse points underneath a terminal bloom.

Dichotomous

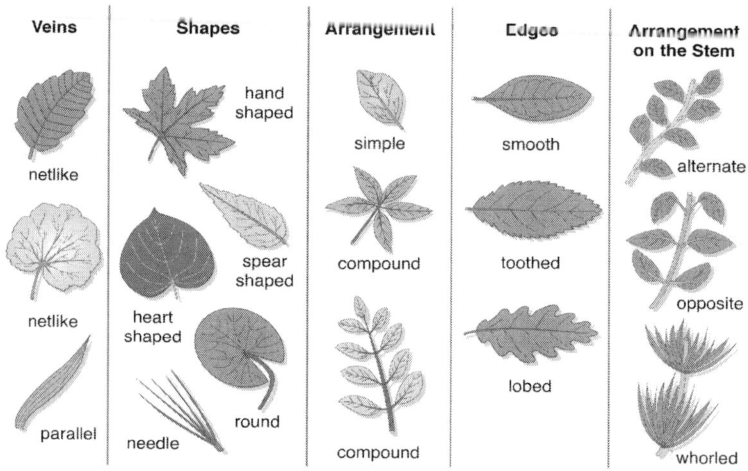

Figure 54: The dichotomous key is used to differentiate different types of leaves.

Source: https://kids.britannica.com/students/article/dichotomous-key/624999

A term that defines separated or isolating into two sections or characterizations. In plant science, it portrays the kind of plant branching that comes when the developing point (apical bud) partitions into two equivalent growth point, which thus partitions in a comparative manner after a time of development, etc. Dichotomous branching is typical in mosses and ferns.

For angiosperm, in dichotomous branching, these branches develop because of an equivalent division of a terminal bud (that is., a bud developed at the top of a stem) into two equivalent branches that are not gotten from axillary buds, albeit axillary buds are available somewhere else on the plant body. A couple of examples of dichotomous branching among angiosperms are tracked down just in certain desert flora, palms (Arecaceae), and paradise bird plants (Strelitziaceae).

The two methods of axillary branching in angiosperms are sympodial and monopodial. Monopodial spreading happens when the terminal bud keeps on developing as a focal pioneer shoot and the horizontal branches remain subordinate—for example, beech trees (Fagus; Fagaceae). Sympodial expansion happens whenever the terminal bud stops to develop (ordinarily when the terminal blossom has developed) and a secondary bud or buds become fresh pioneer shoots, called restoration shoots—e.g., the Joshua tree or scientifically *Yucca brevifolia*; Asparagaceae. Plants with monopodial development are typically pyramidal in shape, whereas those with sympodial development regularly look like a candelabra.

Dicotyledon

Dicotyledon, simply dicot, includes any part of the blossoming plants, or angiosperms, with a couple of leaves, or cotyledons, within the incipient organism of the seed. There are around 175,000 known types of dicots. Most typical nursery plants, bushes and trees, and wide-leafed blooming plants like roses, magnolias, roses and hollyhocks are dicots. Dicots normally have bloom parts (petals, sepals, stamens, and pistils) in view of a structure of four or five, or products thereof, despite the fact that there are special cases (Taiz et al., 2015).

The leaves are net-laced mostly, which implies the vessels that lead water and food have a meshlike design. In the stems, the vessels are generally organized in a nonstop ring close to the stem surface. Around 50% of all dicot varieties are woody; they display a yearly expansion in stem distance across because of the creation of new tissue by the cambium, a layer of cells that stay fitted for division for the duration of the existence of these plants. Spreading of stems is normal, as are taproots. The tiny pores (stomates) on the leaf structure are typically dispersed and are in different directions. The pollen grains ordinarily have three germinal wrinkles or pores (tricolpate status), besides in the more primal families.

Digitate

A palmately compound leaf comprising of six leaflets. No buds exist between the leaflets except for there would be a bud close to the top of the lower aril. These characters allude to the 6th leaf stage. The 6th leaf is the 6th genuine leaf delivered later the cotyledons. On seedlings with a substitute or spirally organized leaves, it is basically the 10th leaf delivered. On seedlings with inverse leaves, it is one of the leaves in the third pair of genuine leaves delivered. On seedlings with whorled leaves it will be one of the leaves in the proper whorl; for example, assuming that there are two leaves in every whorl then the 6th leaf will be among the leaves in the primary whorl.

Dimorphic

A minor subset of parasites inside the Ascomycota phylum is called dimorphic, which alludes to the ability to change over between two different morphologic structures, yeast and hyphae. The thermally dimorphic organisms are special among parasitic microorganisms since they can contaminate people with common and impaired immune protections. A few types of dimorphic fungi are key microorganisms of people and different organisms, including *Candida albicans*, *Blastomyces dermatitidis*, *Coccidioides immitis*, *Paracoccidioides brasiliensis*, *Histoplasma capsulatum*, *Emmonsia* sp. And *Sporothrix schenckii* (Szpak, 2014).

Dimorphic fungi are organisms that can switch between two morphologies during their lifespan: yeast and hyphae. In warm dimorphic fungi, morphologic changes can be incited by temperature. This group consists of significant human microorganisms for the genera Sporothrix, Talaromyces, Emergomyces, Blastomyces, Coccidioides, Histoplasma, Paracoccidioides. They generally have a natural habitat in soil and in plants where they develop as a mold, and when contagious propagules are breathed in or injected by injury in a mammalian host, they go through a complicated interaction and convert into pathogenic yeasts, causing intense mycoses that are typically endemic to certain topographical regions. This morphological progress is critical to pathogenicity.

Dioecious

(Particularly of plants) having both the male and female parts in different plants; having separate genders. Just the female plant produces fruit. Dioecious plants comprise of the male and female blossoms on various plants. So not exclusively does the plant have separate male/female blossoms, they have male plants (with just male blossoms) and female plants (with just female blossoms). Asparagus and hollies are dioecious. Since just the female plants can create fruit, hollies should have a male plant and a female plant adjacent. Male holly plants are frequently given manly names like 'Jim Dandy', 'Blue Prince', or 'Southern Gentleman', so they are not difficult to perceive. Typically, a couple of male hollies are regularly tucked behind the female hollies to guarantee fertilization and fruit arrangement and to conceal the male plants which don't deliver the pompous natural product.

The male cultivars of asparagus are more famous with nursery workers than the females. This is on the grounds that the male lances are bigger, they don't squander any space on fruit production. Male plants can likewise be neater (Stenberg, 2017). Male dioecious trees don't create the "kayak paddle-like" natural products like the females. Male trees of Kentucky Coffeetree, Cork Tree, and Ginkgo are famous for a similar case - no muddled fruit! Some other notable Dioecious plants are; Sago, Mulberry, Ginkgo, Mistletoe, Spinach, Juniper trees, Papaya, Yam, Hemp, Currant Bushes, Hop, Willow, Kiwifruit, and Poplar.

Diploid

Figure 55: Comparison of diploid, tetraploid and octoploid plant cell extracts.

Source: https://medium.com/plant-cell-extracts/up-sizing-the-tale-of-the-poly-ploid-giants-508f9f7db

Diploid defines a cell with duplicates of every chromosome. Practically every one of the cells in the human body carries two homologous, or comparable, duplicates of every chromosome. The main exemption is cells within the germ line, which proceed to deliver gametes, or egg and sperm cells. Microbe line cells are haploid, which implies they contain a solitary arrangement of chromosomes. In diploid cells, one bunch of chromosomes is acquired from the single mom, while the second is acquired from the dad. Individuals have 46 chromosomes in every diploid cell (Singh,

2015). Among those, there are two gender-deciding chromosomes, and 22 sets of autosomal, or non-sex, chromosomes. The complete number of chromosomes in diploid cells is portrayed as 2n, which is double the amount of chromosomes in haploid cells (n).

Diplostemonous

Diplostemonous stamens. In this form of staminal plan, there are many stamens as there are sepals in addition to petals, the stamens exist in two whorls, and the stamens in the external whorl (in green) are each positioned inverse a sepal. With two pairs of stamens; those of the external series embedded inverse the sepals, those of the inward series embedded inverse the petals. Diplostemony is often challenging to define in blossoms at anthesis yet becomes evident with microtome and SEM research of bud improvement of obdiplostemony which is likewise material to diplostemony

Discolorous

Of an alternate or different shading from a neighboring or encompassing part. Botany. Of various colors; having portions of diverse shadings, particolored; particularly (of a leaf) having the upper and lower surfaces of distinct color tones.

Disjunct

In plant science, a taxon with a disjunct dispersion is one whereby two groups are connected yet significantly isolated from one another topographically. The causes are shifted and may exhibit either the development or withdrawal of plants species range (Schuurink et al., 2006). Disjunct dispersions might be brought about by environmental changes, for example, mountain development and continental float or rising ocean levels; it might likewise be because of a plant growing its reach into new regions, by such means as boating, or different animals dispersing plant organisms to another area (sow seeds devoured by birds and organisms can be transported to new areas during bird or animal relocations, and those seeds can be planted in new areas in fecal content.

Different conditions that can deliver disjunct dispersals include: flooding, and wind changes, stream, and current streams, in addition to others, for example, anthropogenic presentation of external presented species either coincidentally or purposely (farming and cultivation). Habitat fragmentation: Disjunct distributions can happen when the environment is divided, which produces divided populaces, and when that fracture turns out to be dissimilar that species development between one reasonable territory to the following is upset, a secluded populace can be created. Extinctions actually cause disjunct dissemination, particularly in regions where just dispersed regions are livable by animal categories; for example, island chains or particular heights along a mountain range or regions along a coast or between waterways like ponds, streams, and lakes. There are various examples of disjunct disseminations at diverse scales including: Europe - East Asia, Europe-South Africa (for example genus Erica), Irano-Turanian disjunction, Mediterranean-Hoggart disjunction (genus Olea), and so forth.

Disk floret

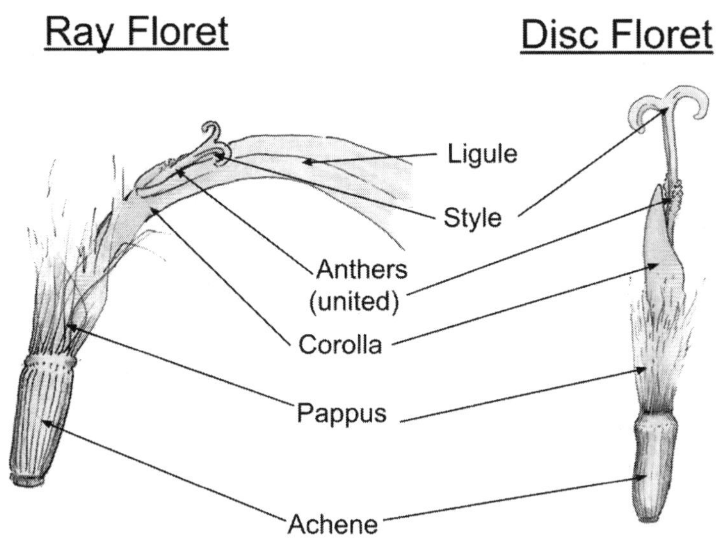

Figure 56: The disk floret stands vertical and compact compared to the ray floret.

Source: http://whatsgrowingincolorado.blogspot.com/2017/01/aster-i-hardly-know-er.html

Among the rounded blossoms in the circle of a composite plant. — Likewise known as disk floret.

The most self-evident and extraordinary general component of Asteraceae is that the blossoms are assembled naturally into dense inflorescences (heads) that cursorily look like individual blossoms. Each such head is customarily subtended by an involucre of little modified leaves (bracts).

Moreover, in the greater part the individuals from the family, the blossoms in the peripheral line or columns of the head have a changed, fundamentally level and lengthen corolla that takes after a singular petal of most different blossoms (Smith & Knapp, 2001). Consequently, the "petals" of a daisy or sunflower are really the peripheral blossoms of the head. An inflorescence of this family can have in excess of 1,000 individual blossoms (florets), and the heads can be gathered into unique, auxiliary plans called capitulescences.

When present, the rose petals of Asteraceae are combined by their edges, developing a cylindrical or tie formed corolla (a sympetalous corolla) that frequently has apical teeth constituting the petal tips. The other flower parts are appended to the highest point of the ovary rather than below it. The Asteraceae calyx (sepals) are so extremely modified, as opposed to that of different families, that it is given an alternative name, the pappus. The pappus comprises of multiple dry scales, awns (tiny pointed cycles), or slender (hairlike) bristles; in some, the scales might be joined by their edges to form a crownlike ring at the culmination of the ovary. For some genera (e.g., Marshallia) the calyx comprises of five consistently positioned scales that are clearly homologous with sepals.

Dissepiment

Dissepiment, "the portions in a fruit brought about by the grip of the margins of carpellary leaves"; a partitioning section, as that amongst the polypore pores; a divider or segment between neighboring locules of an ovary, pericarp or organic product, "brought about by the bond of the edges of carpellary leaves.

Distal

Distal means the piece of the plant which is farthest from the initial point of connection for example the apex of the leaf, shoot or root, whereas proximal means closest to the place of connection. The distal tip of either the shoot or root is that farthest from the stem-root intersection of the plant and closest to the apex of the root or shoot.

Distyly

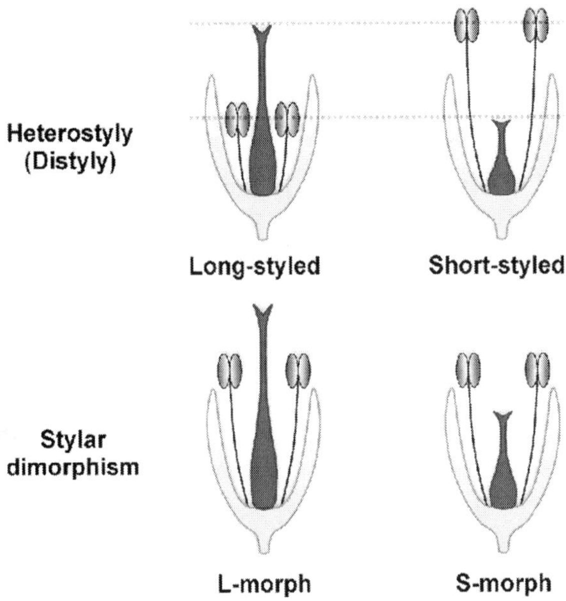

Figure 57: Distyly is a type of heterostyly where the plant needs separate male and female flower parts to propagate.

Source: https://www.researchgate.net/figure/Heterostyly-specifically-distyly-and-stylar-dimorphism-displaying-whorls-arrangement-of_fig5_327685131

Distyly is a form of heterostyly where a plant shows proportional herkogamy. Most distylous plants are self-inconsistent so they can't fertilize ovules within their own blossoms. Clearly, these plants show intra-transform self-incongruence, blossoms of a similar morph are inconsistent. Distyly is a kind of heterostyly in which a plant exhibits equal herkogamy. This propagating structure is portrayed by two separate blossom transformations, where individual plants produce blossoms that either have long styles and short stamens (customarily referred to as "pin", "long-transform" or "L-transform" blossoms) or that have short styles and long stamens (customarily known as "drone", present day terminology calls these as the "short-transform" or "S-transform" flowers). However, distyly can allude to any plant with two morphologies if any one of these qualities between blossoms delivered by various plants is valid; there is a distinction in fiber length, dust size or shape, style length, or the outer layer of the stigma (Simon & Petrášek, 2011). Distylous species that don't display genuine self-incompatibility by and large show an inclination towards cross-morphs - which means they show higher reproduction rates while duplicating with an entity of the inverse morph.

Diurnal

Plants that release their blossoms in the daytime are portrayed as diurnal, whereas those that sprout during evening are nocturnal. The period of bloom opening is frequently identified with the time at which certain pollinators are scavenging. Diurnality is a type of plant behavior defined by movement during the daytime, with a time of inactivity or other idleness being around evening time. The normal descriptive word utilized for daytime movement is "diurnal". The conduction of action by a organism relies upon a number of ecological factors, for example, the temperature, the capacity to accumulate food, the danger of predation, and the season. Diurnality is a pattern of movement inside a 24-hour time frame; cyclic exercises called circadian rhythms are endogenous cycles not reliant upon outer prompts or ecological elements aside from a zeitgeber. For instance, sunflowers open during the day to draw in honey bees, while the night-sprouting cereus opens around evening time to draw in huge sphinx moths.

Domatia

(Plant science) tiny structures on the lower side of a leaf in some wooden dicotyledons, typically comprising of depressions, mostly encased by leaf tissue or hairs, situated in the axils of the main veins. Domatia (single: Domatium) are structures generated by plants for arthropods to utilize. Generally, domatia are tiny chambers inside the plant or made by hairs on leaves (Rusman et al., 2019).

The domatia are expected to be utilized by plants that have an advantageous connection with the plant. For instance, Pseudomyrmex subterranean insects live in domatia inside the thistles of some Acacia trees. The tree benefits from the insects as the insects eat organisms benefiting from the Acacia. Consequently, the subterranean insect habitat is secured inside the domatia. The thistles are empty and form domatia for subterranean insects. The subterranean insects enter the domatia by means of the opening between the two thistles.

Drupe

Drupe, in natural science, fruit product that typically contains a solitary seed, like the cherry, peach, and olive. As a basic organic product, a drupe is gotten from a solitary ovary of a singular bloom. The external layer of the ovary divider is a meager skin or strip, the center layer is thick and typically beefy. However in some cases intense, as in the almond, or sinewy, as in the coconut, and the internal layer, known as the pit, or putamen, is hard and stony. The pit, which is regularly mistaken for the actual seed, as a rule, has one seed or, seldom, a few, in which case just one grows completely. Other minor drupes are mango, pecan, and dogwood.

Drupelet

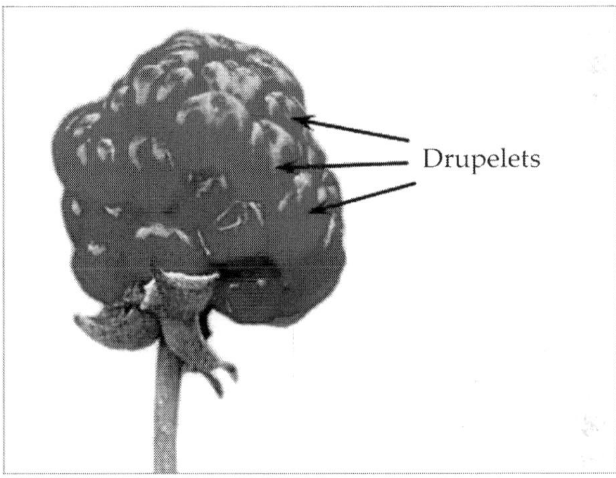

Figure 58: Red raspberry drupelets.

Source: https://southfloridareporter.com/raspberry-100-drupelets-every-drupelet-considered-separate-fruit/

A "drupelet" is the singular seed on natural fruits like raspberries and blackberries. Customarily, cool, muggy air disperses and retains UV radiation, while hot, dry air has the contrary impact and permits more linear UV beams to arrive at the fruit. In some fruits, such as the raspberry and blackberry—which are false berries—numerous little drupes are bunched together (Reid et al., 2009). Officially, these fruit products are called totals of drupelets. Such fruit products are shaped in different combined carpels (ovule-containing structures) from a solitary blossom.

E

Edaphic

Edaphic refers to something that has been influenced by the soil. Various plant species have their growth habits influenced by the nature of soil. A taproot plant growing in dry soil will tend to have a longer taproot. This enables the roots to penetrate into deeper layers allowing them get more water from the soil. Plants growing in soils rich in nutrients and minerals tend to be much healthier and stronger compared to others. Also if there are several pests in the soil, the plant will be weak due to pest infestation among others. Soils influence the likelihood of a plant growing in a certain area.

Elaiosome

An elaiosome is an external structure that is attached to the seed of a plant. In some cases, the elaiosome may look fleshy. In some species of plants, the elaiosome is rich in nutritious materials while in others they are rich in oils. Elaiosomes have different functions that may depend on the species of the plant (Poff & Zimmerman, 2010). The functions however are not always obvious in plants. Elaiosomes are known to attract ants and other animals that may function pollutants allowing the dispersal of pollen allowing fertilization. In other species of plants, elaiosomes may repel certain animals ensuring that they do not eat the seed.

Elephophily

Elephophily is a form of pollination that involves spores or pollen being distributed by the feet of elephants. The Rafflesia is a plant whose flowers are known to exhibit elephophily. The plant is very large and is found at ground level. The flower produces pollen grains that get attached to the feet of elephants. They are then carried to the stigma of another flower. Flowers that experience elephophily are known to have an unpleasant and strong smell similar to decaying flesh. Elephophily flowers are usually unisexual. For pollination to be successful, there is a need for the right proximity between the flowers.

Emarginate

Emarginate refers to the presence of a notch at the apex. In most cases emarginate leaves have their apex notched downwards at the midvein. This is towards the petiole. In the case of the five-leaved Akebia, though is a compound leaf each of the leaflets ate notched. The umbrella pine is emarginate though it is not very obvious unless it is observed on close inspection. This characteristic is used when identifying the plant. In higher plants, the emarginate is considered the main organ of transpiration and photosynthesis.

Embryo

An embryo is a young plant in a seed prior to germination. The embryo is also called the seed embryo. After fertilization, the embryo develops to form an adult plant and flowers. The embryo is generally contained in the bud or seed and its role is to act as a "starter kit" for the plant. For the embryo to develop, there are some conditions that should be met. These conditions activate the embryo and begins the germination process. After germination the embryo becomes a seedling that grows out of the seed. The embryo contains the earliest form of a plant's stem, leaves and roots (Pauchard et al., 2009). They use the seed to sense whether conditions are right to allow germination. These conditions the presence of adequate mineral salts in the environment, adequate water and adequate oxygen. When these nutrients are available, the embryo begins to consume the endosperm while in the seed. This allows the new plant to start growing. After the embryo consumes the endosperm, it bursts out of the seed. It then grows to form the shoots and the root. The shoot grows out of the ground and then develops to form the stem and leaves.

Emergent

Emergent is a term that refers to a plant that is taller than the surrounding vegetation. It could also refer to plants that are taller compared to other species of plants on land. For aquatic plants, emergent plants are those that bear flowers and leaves that are above the water surface. Examples of aquatic plants that are emergent include papyrus, reeds and lilies. The Stuckenia is a pondweed that is not emergent until it flowers or the flowers appear above the water surface. Emergent aquatic plants are rooted in the ground with their leaves, stems and flowers arising above the water. They also rely on aerial reproduction. Their roots enable them to gain nutrients from the soil. Aquatic emergent plants are used as supply cover and provide plants with nesting areas. There are those who consider emergent aquatic plants as a nuisance.

Endemic

Endemic is used to refer to things having a natural distribution that is restricted to a particular geographic region. Plant species can be endemic to either small or large areas of earth. Some plant species are endemic to particular to certain parts of the continent while others are known to grow on a single island (Prajapati & Modi, 2012). In some cases, plant species that are in a given area are isolated creating difficulty for the plants to spread to other areas. Some of these species have developed an adaptation to the area they are growing. The occurrence of endemic plants is more common in certain regions.

Endocarp

Endocarp is the innermost layer of the wall of a fruit. In botany, the term endocarp refers to the inside layer of the pericarp which directly surrounds the seed. For most plants, the endocarp is a membranous structure. For instance, in citrus plants, the endocarp is the only part consumed. In other plants, the endocarp is hard and thick as in the pyrenes of drupe fruits of the Rosaceae family such as apricots, plums, cherries and peaches. In nuts, the endocarp is the Stony layer that surrounds the kernel of walnuts and pecans among others. For this reason, the endocarp is removed before consumption.

In citrus fruits, the endocarp is separated into sections called segments. There are juice vesicles in the segments. These juice vesicles contain the juice of the fruit. The endocarp serves to protect the seed from damage.

Endodermis

Endodermis is the central and innermost layer of the cortex in plants. The endodermis is a cylinder of compact living cells having radial walls with hydrophobic substances impregnating it. The casparian strip restricts the apoplastic flow of water to the inside. The endodermis is found between the stele and cortex. The endodermis is distinctly visible as a layer of cells in seedless vascular plants and is found outside the stele in shoots and roots. Vascular plants with seeds more so woody plants, the endodermis is present in the roots but absent from the stem. The endodermis is useful in regulating the movement of hormones, ions and water into and out of the vascular system. It is at times used in storing starch and protecting plants from toxins that move into the vascular system (Neumann et al., 2009). The endodermis is composed of a single layer of barrel-shaped cells without intercellular spaces. In some cases, several cell layers make up the endodermis. The four sides of the endodermis cells are transversely and racially thickened with suberin. Suberin is a watery impermeable waxy substance that is deposited in casparian strips which are distinctive bands. The casparian strips have variable widths though they are smaller than the cell wall. Often at times, the endodermis is likened to a brick cylinder with the individual cells bring represented by the bricks while the mortar between the bricks is analogous to the casparian strip. Suberin is extensively deposited on all cell wall surfaces in older endodermal cells. In this case, the cells become lignified and form a complete waterproof layer.

Endosperm

Endosperm is a tissue that surrounds the embryo in the seeds of most flowering plants. The endosperm is produced after double fertilization. The endosperm is triploid meaning there are three chromosomes sets in the nucleus. It's is also driven by auxins. While surrounding the embryo, the endosperm provides nutrients to the embryo in form of starch. In some cases, the endosperm can contain protein and oils. This makes the endosperm a good source of nutrients to animals. In plants like wheat, the endosperm is grounded to make flour for bread while in barely it is used as the main source of sugar when producing beer. In coconut, the endosperm forms the edible portion. There are certain plants whose seeds do not have endosperms. They include orchids.

Endospory

Endospory is the production of spores which after germination become a reduced multicellular gametophyte contained within the spore wall. The term could also be used to refer to the partial or entire retention and development of gametophytes within the wall of the generative spore. Endospory is mostly noted in heterosporous plant species. Endospory provides some ecological advantages to plants in that as the gametophyte develops within spore walls gives them improvement in sexual function and also protection from harsh conditions (Nikunj et al., 2014). The spore wall also supplies the nutrients required by gametophytes during growth. This results in endosporic megagametophytes with an increased ability to store metabolites. This increases the time spores are able to live without water while increasing their ability to populate new habitats. It is the opposite of exospory.

Entomophily

Entomophily also called insect pollination is a form of pollination whereby insects are involved in the pollination of both flowering and non-flowering plants. Flowering plants are able to attract insects as they have bright colored petals and conspicuous patterns that lead to rewards of nectar and pollen. In addition to that, flowering plants have attractive scents that mimic insect pheromones. Need and other insect pollinators have adapted to enable them to pollinate plants. Some of the adaptations include the presence of pollen baskets at their hind legs and sucking or lapping mouthparts to take in nectar (Novello et al., 2020). The development of pollination behavior came as a result of the required coevolution of flowering plants and insects by pollination and insect mechanisms by the flower. Flowering plants are suitably adapted to enable pollinators to take pollen from the flower. The filament of the male flower tends to be very strong enabling it to support the weight of the pollinator.

Epecophyte

Epecophytes are species of recent appearance, constant and numerous in the country though they are confined to artificial habitats. Ruderal vegetation and meadows are examples of epecophytes as they are dependent on humans for their existence. These plants may require certain conditions to be met to ensure their survival.

Epicalyx

Epicalyx is a ring of small bracts at the base of flowers. The epicalyx resembles an extra outer calyx. It forms an additional whorl around the calyx of a single flower. The epicalyx is a modification of bracteoles. The epicalyx can also be defined as a group of bracts that resemble a calyx or as bracteoles that form a whorl outer to the calyx. As a calyx-like extra whorl of floral appendages, each segment of the epicalyx is called an episepal. This is because they resemble sepals in them. Epicalyx is present in the Malvaceae family. However, plants like strawberries may or may not have an epicalyx.

Epicarp

Epicarp is a term used in botany to refer to the outermost layer of the pericarp. For most fruits, the epicarp forms the tough outer skin of the fruit. In some cases, the epicarp is also called the exocarp or generally as skin. In citrus fruits, the epicarp is also called flavedo. The epicarp consists of a layer of elongated living cells called the epidermis. The epidermis is also covered by a thin film of the cuticle (Mercuri et al., 2002). The epicarp protects underlying tissues and permits the exchange of gases and metabolites with the external environment through lenticels. Wax is deposited on the cuticles which forms a light gray blood that gives the skin a dull appearance. The wax also makes the skin impermeable to water.

Epicormic

Epicormic is a term used to refer to flowers, shoots and buds developing from the old wood of trees. With regards to the shoot, the epicormic shot is a shoot that grows from an epicormic bud. The epicormic bud lies underneath the bark of a branch, stem or trunk of plants. Epicormic buds are dominant beneath the bark as growth is suppressed by hormones from active shoots. Epicormic buds develop into active shoots under certain conditions. These conditions include damage to the higher parts of the plant or when there is an increase in light after the removal of nearby plants. Epicormic shoots and buds occur in many woody species but are absent in others. Conifers have epicormic shoots and buds. Trees regrow after pollarding and coppicing by means of epicormic shoots. This includes branches and tree trunks being cut back on a regular cycle. Forestry techniques are applied in animals that possess growth abilities.

Epicotyl

Epicotyl is the region of a seedling stem above the stalk f the seed leaves of an embryo. It is between the first foliage leaves and cotyledonary nodes. The epicotyl exhibits rapid growth showing hypogeal germination and extends the stems above the soil surface. There a number of misconceptions about the epicotyl including the notion that as the epicotyl is closer to the apex is the first part to emerge after germination. During hypogeal germination, the hypocotyl forms a hook and pushes out of the soil. This allows more delicate tissues of the apical meristem and plumules to avoid damage through pushing through the soil.

The epicotyl expands from the point of attachment of the leaf primordia and the shoot apex. While growing the cotyledons may remain below ground or it may be pushed up aboveground with the growing stem. All of this depends on the species of plant (Matyjaszczyk & Śmiechowska, 2019).

According to plant physiology, epicotyl is an embryonic shoot above the cotyledon. For this reason, the epicotyl develops into the leaves of the plant. In monocots, the epicotyl is the first shoot that emerges from the ground. For the epicotyl, the first shoots and leaves emerge. In dicots, the base stem under the spent withered cotyledons is the hypocotyl while the epicotyl is the shoot above. The phytochrome photoreceptors control the lengthening of the epicotyl.

Epicuticular wax

Epicuticular wax is a wax coating covering the outer surface plant cuticles in land plants. Epicuticular wax may form a bloom or whitish film on fruits, leaves and other plant organs. The chemical composition includes the presence of hydrophobic organic compounds. The organic compounds are mainly straight-chain aliphatic hydrocarbon that may or may not have substituted functional groups. The epicuticular wax is very important in plants as it serves to decrease surface wetting and moisture loss. They also reflect ultraviolet light, act as an anti- climb surface and assist in the formation of an ultra hydrophobic and self-cleaning surface.

Epidermis

Epidermis is a single layer of cells that form the cover for stem, roots, flowers and leaves. The epidermis forms a boundary between the external environment and the plant. The epidermis performs several functions in plants. They include providing protection against water loss, absorption of water and mineral salts, secretion of metabolic compounds and regulation of gas exchange. A dorsoventral anatomy is noted in the epidermis of most leaves. There is a difference in the construction in the abaxial and adaxial surfaces and therefore they serve different functions. In woody stems, a secondary covering is produced that replaces the epidermis by functioning as a protective covering (Mandel, 2019).

Epigynous

Epigynous is a term used to define flowers whose ovary is inferior to the other parts. The ovary is enclosed inside the thalamus and other three whorls are located above it. In epigynous flowers, the stamens, petals and sepals are seen to arise from the top of the ovary. This makes other floral whorls to be superior while the ovary is inferior and observed below the flower. In epigynous flowers, the thalamus is flask-shaped. Examples of epigynous plants include daffodils, cucumber and the common honeysuckle. Compared to hypogynous flowers, epigynous flowers occupy a higher rank. All flowers of the composite family are epigynous.

Epilithic

Epilithic is a term used to refer to plants that grow on the surface of a rock. Examples of epilithic plants include ferns and mosses. These plants are able to gather all the required nutrients from small amounts of organic matter or rainwater that accumulate on rock surfaces. Chasmophytes are examples of plants that grow specifically on cracks on rocks. The term epilithic is derived from the joining of two terms; "epi" which means upon and "lithos" which means stone (Lindström et al., 2013).

Epinecral

Epinecral is a layer of horny dead fungal hyphae with indistinct lumina or a layer that is found near the cortex above the algal layer. In lichens, the epinecral is a layer of dead tissue found above the surface of the cortex of a lichen.

Epipetalous

Epipetalous is a condition in flowers whereby the stamens are attached to the petals of the flower instead of being directly over the thalamus. Solanum is an example of an epipetalous plant. For lilies, their petals and sepals are the same and called perianth. When the anther is attached to the perianth is called epiphyllous.

Epipetric

Epipetric is a term used to refer to plants that grow on rocks. It is one of the classifications of lithophtytes. Epipetric plants are also called epilithic plants and mosses are examples of epipetric plants. These plants are able to gather all the required nutrients from small amounts of organic matter or rainwater that accumulate on rock surfaces. Epipetric plants can also be classified as either being facultative or obligated. Obligate epipetric plants are those that grow slowly on rocks while facultative are those that grow partially on rock and another substrate (Luan et al., 2020).

Epiphloedal

Epiphloedal are plants that grow on the surface of the bark. Epiphloedal plants are the opposite of endophloedal plants that grow inside the epilithic plants which grow on rocks and not bark. Epiphloedal plants are mostly found on woody plants as they are known to have barks. Epiphloedal plants may draw their nutrients from the plant.

Epizoochory

Epizoochory is a type is seed dispersal known to occur when fruits or seeds physically adhere to the outside of vertebrates. The seed is therefore transported on the outside of animals more so those that have hairs and fur. This form of seed dispersal mostly occurs in plants whose seeds are to be transported to far destinations to avoid competition and increase the chances of the new plant surviving (Lambers et al., 2008). The act of plants using animals when dispersing their seeds is called zoochory. When the seeds stick to the surface of the animal it is referred to as epizoochory. The term "epi" means upon. Epizoochory is noticeable when burrs stick on dog's fur. Epizoochory is an adaptation in plants that allows them to spread their progeny. Voles, raccoons, foxes and squirrels are used in the dispersal of seeds.

Epruinose

Epruinose in botany refers to plants that are not pruinose. Pruinose plants are those that are covered with a powdery waxy material or having a bloom. Therefore epruinose is a plant that is not covered with a powdery waxy material.

Ericoid

Ericoid refers to leaves that resemble those of the European heaths in that they are small and sharply pointed. The term ericoid is used to conveniently describe small, tough leaves. When referring to plants, the term ericoid means that apart from its sclerophyllous leaves. The plants have short internodes so that the leaves may cover the slender branchlets.

Ethelochoric

Ethelochoric refers to the deliberate introduction of plants, seeds and seedlings by human beings into a new habitat. It is a form of hemerochory as it involves the distribution of seed, cuttings or plants by human beings into areas that the plant could not colonize through their natural mechanisms of spread. After the plant is introduced to the new habitat, it is able to sustain itself without any form of help from humans. Ethelochoric plants can either increase or decrease the biodiversity of a habitat. The deliberate introduction of plants has been useful in improving forest cover (Klassen, 2001).

Even-pinnate

Even-pinnate refers to a leaf shape characterized by a pair of leaflets at the apex. Even-pinnate is also called paripinnate where by pinnately compound leaves with leaflets borne in pairs along the rachis without a single terminal leaflet. It is therefore a compound leaf composed of more than one part as it has several lobes. Even-pinnate leaves have even numbers of leaflets on the leaf.

Evergreen

Evergreen is a term used in botany to refer to plants having foliage that remains green and functional through more than one growing season. The term is also used to refer to plants that retain their foliage only in warm climates. Evergreen plants are different from deciduous plants. Deciduous plants are known to completely lose their foliage during the dry or winter season while evergreen plants retain their foliage. Examples of evergreen plants include Clubmosses, most angiosperms from frost-free climates such as rainforest and eucalyptus, holly, live oak and gymnosperms such as cycads (Katova, 2007). Most species of conifers such as red cedar, blue spruce, hemlock and pine are evergreen plants. Even though evergreen trees do not lose their leaves, there are some leaves that are gradually lost but at different times. Most tropical rainforests are considered to be evergreens. They replace their leaves as they age and other leaves are lost. The warm temperate plants are also evergreen. Very few plants in cool temperate climates are evergreen.

Exalbuminous

Exalbuminous is a term used to refer to seeds of a given species that have no endosperm. They may not have an albumen. Radish, sunflower, squash, walnut, oak, pea and bean are examples of exalbuminous seeds. Dicots such as castor bean and Brazil nu while monocots such as palms and grasses have albuminous seeds. All gymnosperms are albuminous.

Exocarp

Exocarp is the outer layer of the pericarp and forms the skin of fleshy fruits. As the outer layer of the pericarp, the exocarp plays different roles which include dispersal or protection. The exocarp is also called the husk, peel, rind or skin of a fruit. The exocarp covers the mesocarp which is then followed by the endocarp. For fleshly fruits, the endocarp has bright colors that help attract other animals which eat them hence causing the dispersal of the seed to new areas after consumption. The exocarp also protects the mesocarp from external damage by acting as the outer covering (Jeger et al., 2007).

Exospory

Exospory is the production of spores that germinate into free-living gametophytes that are multicellular in nature. Exospory is the opposite of endospory. Exospory results in the production of exospores which are formed on either one end of the vegetative part of the cell. This means that they are formed on the surface of the vegetative cell. The production of spores allows the reproduction of these plants ensuring their continuity and survival.

Exotesta

Exotesta is the outer layer of the tests from the outer surface of the outer integument. It is located close to the endotest which is the inner epidermis of the outer integument. In some seeds, the exotesta functions as a mechanical layer and therefore the seed is referred to as exotestal seed. However if the endotegmen forms the mechanical layer the seed is then referred to as endotestal. The exotesta consists of one or more rows of cells elongated and are palisade like. This is common in Fabaceae and is therefore called palisade exotesta. As a mechanical layer, the exotesta functions as a protective layer. It protects the seed from mechanical damage. It also helps provide information to the embryo on the external environment. This enables the seed to adapt to the environment. It can also trigger germination in plants.

Exotic

Exotic refers to a plant that is not native to a certain region. Exotic plants are those that have been introduced from another region or country. Other than plants, the term could also be used to refer to flowers. Exotic plants and flowers could also be referred to as non-native or foreign plants and flowers (Hirschberg, 2001). After these plants are introduced to a specific region the species become dominant and invasive over that region. Exotic plants are thought to cause imminent damage to neighboring plants and their environment.

F

F1 Hybrid

The term F1 hybrid seeds refers to a plant that has been selectively bred by cross-pollination two separate parent plants. The phrase is an acronym for Filial 1, which literally means "first children" in genetics. It is frequently written as F1, however, the terms are interchangeable. Hybridization has been around for quite some time. In the nineteenth century, Gregor Mendel, an Augustinian monk, first documented his achievements in cross-breeding peas. He cross-pollinated two different yet pure (homozygous or same gene) strains by hand. He saw that the plants that grew from the subsequent F1 seeds had a heterozygous or divergent gene make-up. These new F1 plants possessed traits that were dominant in both parents but were not identical to either. The peas were the first reported F1 plants, and Mendel's research gave birth to the study of genetics.

Figure 59: Much is written in today's gardening community about the desirability of heirloom plant varieties over F1 plants. What are F1 hybrid seeds? How did they come about and what are their strengths and weaknesses in the home garden of today?

Source: https://www.gardeningknowhow.com/garden-how-to/propagation/seeds/f1-hybrid-seeds.htm

F2 Hybrid

The second filial generation; hybrids resulting from self-pollination within a population of F1 hybrid plants (Hadwiger, 2013).

This is the product of two F1 plants that have been crossed. This is considered the second generation. This will not necessarily produce a great plant.

Fabiform.

Shaped like a kidney bean.

Fasciculate

Facultative means "optional" or "discretionary" (opposed to obligate), and is mostly used in biology in words like:

A facultative photoperiodic plant is one that will eventually flower regardless of night duration but is more likely to flower in proper light conditions (Guo & Liu, 2007).

Photoperiodism is the physiological response of an organism to the length of the night or a dark period. It's found in plants, animals, and fruit.

Facultative saprophyte, lives on dying, rather than dead, plant material

Female Flower

Flowers are plant parts.

Flowers are not only beautiful, but they also play a crucial role in the formation of seeds. Flowers are made up of several basic components. The pistil is the feminine portion of the plant. The pistil is often found in the middle of the flower and consists of three parts: the stigma, style, and ovary. The stigma is a sticky knob located near the top of the pistil. It is attached to the style, which is a long, tube-like structure. The style leads to the ovary, which contains female egg cells known as ovules.

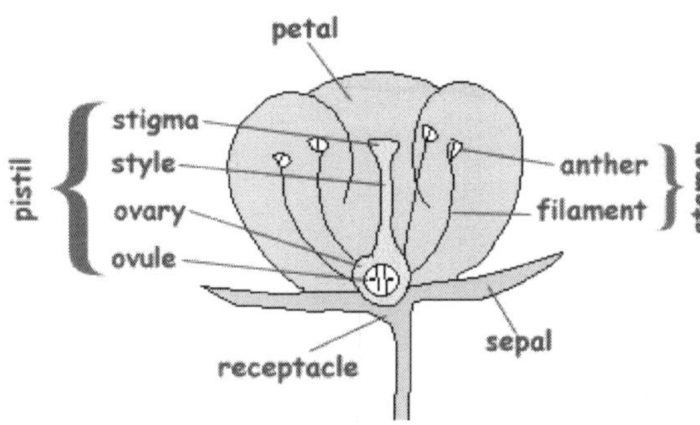

Figure 59: Flower anatomy.

Source: https://web.extension.illinois.edu/gpe/case1/c1facts2d.html

The male parts are called as stamens, and they frequently surround the pistil. The stamen is made up of two parts: an anther and a filament. The anther produces pollen (male reproductive cells). The filament is what holds the anther together.

During fertilization, pollen settles on the stigma, and a tube stretches down the style and into the ovary. Male reproductive cells go down the tube and fertilize the ovule. The fertilized ovule develops into the seed, while the ovary develops into the fruit.

Fertile

Agronomists, or those who study plant usage, use the term to refer to soil. Plants grow quickly in fertile soil because it includes a high concentration of nutrients. These nutrients, which aid in plant health, are derived from minerals as well as decaying plant and animal debris. Minerals may have been deposited in the soil during the last glacial period, sometimes known as the Ice Age (Craigie, 2011).

Fertilizers can be put into the soil to improve its fertility. Phosphorus, nitrogen, and potassium are among the nutrients included in fertilizers. Humus, or composted plant debris, is a natural fertilizer that can improve soil fertility. Manure, as well as the droppings of certain animals such as bats and cows, is a good fertilizer. Many companies produce fertilizers for certain crops. These fertilizers can be applied to the soil or to the plants directly.

Poor farming practices, certain grazing activities, and erosion can all reduce soil fertility.

Figure 60: A rain forest ridges a patchwork of sugar cane fields near Gordonvale, Australia. Gordonvale is in the state of Queensland, one of the most agriculturally productive areas in the country. In addition to cane, farms in Queensland are able to grow tropical fruit such as bananas and pineapples.

Source: https://www.nationalgeographic.org/encyclopedia/fertility/

Planting the same crop year after year can deplete the soil of nutrients. Overgrazing by too many cows, goats, or sheep can limit the growth of new grasses. Erosion causes soil to be blown away by the wind or drained away by water (Clemens et al., 2013). Fertile soil is typically found in river valleys or areas where glaciers deposited minerals during the last Ice Age. Mountains are frequently less fruitful than valleys and plains. The Pampas, for example, is a part of South America's lush plains. Argentina, Brazil, and Uruguay are all part of the Pampas. Farming and ranching are both possible in the Pampas.

Fertilization

Figure 61: Bee pollinating.

Source: https://byjus.com/biology/fertilization-in-plants/

Fertilization is the process of sexual plant reproduction that happens after pollination and germination.

The joining of male and female gametes (pollen and ovum) to produce a diploid zygote is referred to as fertilization. It is a physicochemical reaction that occurs after carpel pollination. The entire procedure takes place in the zygote as it grows into a seed.

Flowers, as angiosperm reproductive structures, serve a crucial role in fertilization (flowering plants). Plants fertilize themselves when haploid gametes combine to produce a diploid zygote.

Male gametes are delivered into female reproductive organs during fertilization by pollinators (honey bees, birds, bats, butterflies, flower beetles), and the end result is the creation of an embryo in a seed.

Fiber

Fibers are all sclerenchyma cells that form part of the plant's skeleton. They are usually lengthy, thick-walled cells with small pores and pointy ends. Lignin and cellulose are common components of the walls. Fibers can exist alone or in small groups, but they are more likely to build sheets of tissue with overlapping and interconnecting individual cells.

Fibers can be found in almost any part of a plant, including the stems, leaves, fruits, seeds, and so on. The four basic variations include bast fibers, wood fibers, and sclerenchyma cells associated with vascular bundle strands in leaves, and surface fibers, which are hair-like expansions on plant seeds. The term "bast fiber" has been chastised because it does not describe the type of tissue or region where the fibers are found (Franklin et al., 2013).

Cortical fibers, pericyclic fibers, and phloem fibers are possible names for the fibers found in the outer sections of the stem. However, "bast" is a phrase that has been around for a long time and is so well-known in business that it will be utilized in this debate.

Fibers of economic value can be found in a wide range of plant families, particularly those from the tropics.

Fiber Cell

Fibers are significantly enlarged cells with long, curved ends that connect to provide the greatest amount of support to a plant. They are often found in bundles or strands and can be found almost everywhere in the plant body, such as the stem, roots, and vascular tissue in leaves. Many of these fibers, including seed hairs, leaf fibers, and bast fibers, are important raw resources for textiles and other woven goods (Foxcroft et al., 2017).

Sclereids have a wide range of shapes and can be found in a variety of plant tissues, including the periderm, cortex, pith, xylem, and phloem. They can also be found in leaves and fruits, as well as the hard shell of nuts and the outer hard coat of many seeds. Sclereids, commonly known as stone cells, are responsible for the rough texture of pears and guavas.

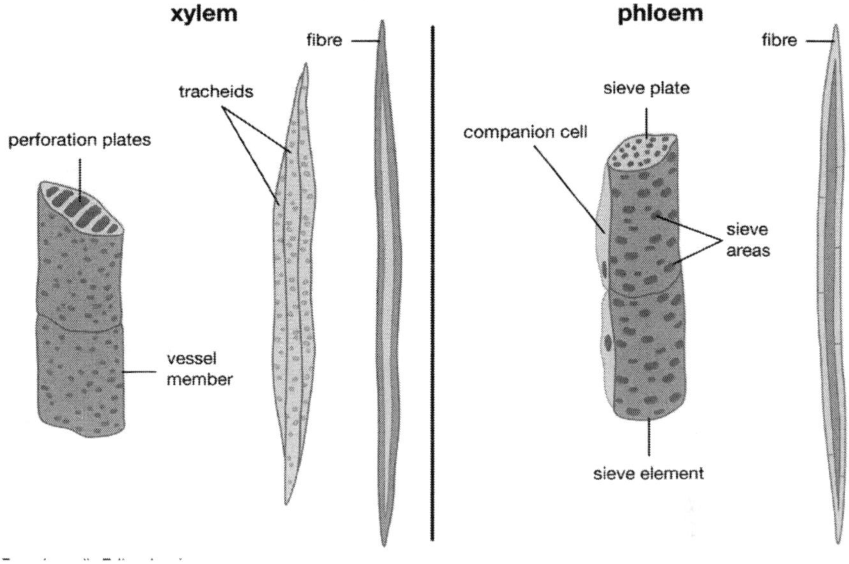

Figure 62: Cells of the xylem and phloem.

Source: https://www.britannica.com/science/fiber-plant-anatomy

Filament

Botanists define filament as the thread-like stalk that supports the plant's anther. A filament and an anther combine to form a stamen, which is the male reproductive component of a flower. Anthers produce pollen, and filaments help hold anthers out to make it easier for wind and insects to spread pollen grains, assisting in species reproduction.

Filaments vary in length as well, even within the same flower. Didynamous stamens are seen in species with two long and two short stamens. Tetradynamous stamens are four long and two short stamens in a flower.

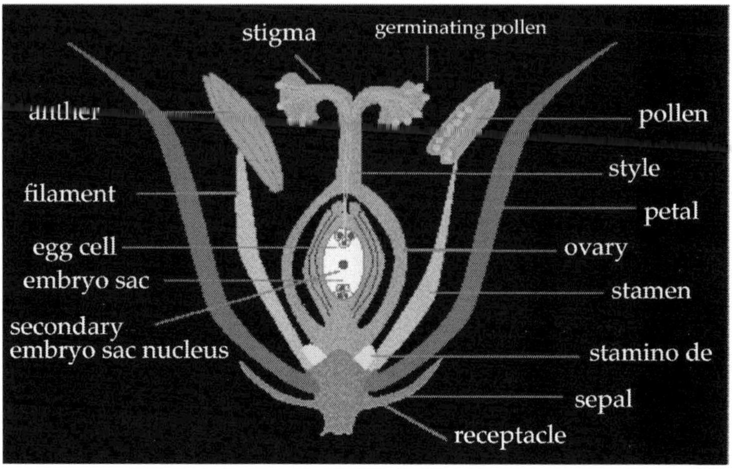

Figure 63: Diagram (cross-section) of a flower.

Source: http://www1.biologie.uni-hamburg.de/b-online/e02/02d.htm

Filamentous

- (Science: cell biology) in the form of very long rods, many times longer than wide.
- consisting of filaments or fibers, hairlike.

Filiform.

- Thread-like. E.g. stamen filaments, or leaf shapes.

Flabellate

Fan-shaped; i.e., broadly wedge-shaped.

Gingko biloba flabellate (flabelliform) leaves and short stems from an unvouchered plant growing at the Botanical Garden Metro North railroad station. It's worth noting that the leaves are flabellate and have small branches from which the leaves emerge (Fahad et al., 2017).

Flaccid

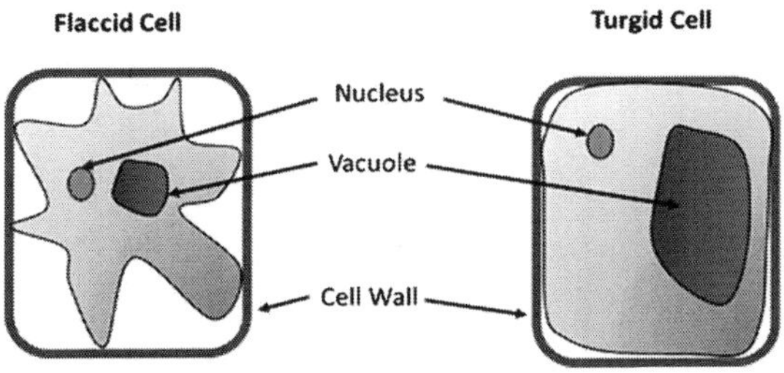

Figure 64: Examples of flaccid and turgid plant cells.

Source: https://www.biologyonline.com/dictionary/flaccid

In botany, a flaccid cell is one that lacks turgidity, i.e. it is not bloated and plump, but rather loose or floppy, and the cell has become drawn in and pulled away from the cell wall.

- (general) weak; soft; lax; lacking vigor; (2) (botany) of a plant cell in an isotonic solution such that the plasma membrane is not forced hard against the cell wall and is thus neither inflated (turgid) nor plasmolyzed.

Flexistyly

Flexistyly is a sexual dimorphism distinguished by two morphs that vary in the time of sexual activity expression and also exhibit reciprocal motion of the stigmatic membrane along a vertical axis throughout the flowering period. The adaptive significance of flexistyly has been seen as a floral tool for outcrossing, but it may also function to reduce sexual disruption in which styles and stigmas hinder pollen export (Elmer & White, 2018).

Floccose

Covered with tufts of soft woolly hairs that usually rub off readily

Flora

The term "flora" refers to both the plants that flourish in a certain location and the dissemination of scientific accounts of those plants. To distinguish between the two, the word is typically capitalized when referring to a publication. A Flora can be anything from a quick summary of plants found in a specific place to a detailed description of those plants. Floras vary from popular textbooks in that they attempt to cover all plants, not just the most common or apparent.

AFlora always includes scientific and common names, literary references, descriptions, habitats, geographic dispersion, photos, flowering dates, and remarks. Specialist information like as plant chemistry, reproduction, chromosome counts, and population occurrences are less typically included in floras (De Waele & Elsen, 2007).

The plants are sometimes listed alphabetically, and other times they are depicted within a classification scheme that identifies which plants are most comparable or are assumed to be related. Floras frequently include "keys" that allow the user to recognize an unknown species.

Floral Leaves

The upper leaves at the base of the flowering branches.

Floral Diagram

Figure 65: *Anagallis arvensis* floral diagram. The primary axis is represented by the dot, and the green structure beneath it is the subtending bract. The calyx (green arcs) is made up of five free sepals, while the corolla (red arcs) is made up of five united petals. Hairy filaments connect antepetalous stamens to petals. The ovaries are superior, the placentation is free central, and the ovules are atropous.

Source: https://en.wikipedia.org/wiki/Floral_diagram

A floral diagram is a graphical representation of the structure of a flower. It depicts the quantity of floral organs, as well as their arrangement and fusion. The various components of the flower are represented by their own symbols. Floral diagrams can be used to identify flowers or to better comprehend the evolution of angiosperms. They were first used in the late 1800s and are widely credited to A. W. Eichler (Chen & Yamaguchi, 2002).

They are often utilized in conjunction with the floral formula of that flower in order to understand its morphology.

Floral Formula

A floral formula is a notation used to express the structure of specific types of flowers. Numbers, characters, and various symbols are used in such notations to express important information in a concise manner. They may describe the floral shape of a specific species, or they may be generalized to characterize higher taxa, usually providing ranges of organ numbers. Floral equations are one of two methods of depicting flower structures created in the nineteenth century, the other being floral diagrams. The format of floral formulae varies according to the preferences of specific authors and periods, but they all transmit the same information.

A floral formula is frequently used in conjunction with a floral diagram.

Floral formulations were created during the start of the nineteenth century. Cassel, who created lists of integers to express numbers of components in named whorls, and Martius were the first authors to use them. In his textbook, Grisebach employed 4-integer series to symbolize the four whorls of floral components to define the properties of floral families, expressing the numbers of distinct organs separated by commas and emphasizing fusion. Sachs utilized them in conjunction with floral diagrams, pointing out their advantage of being composed of "regular typography."

Floral Tube

A hypanthium, also known as a floral cup, is a structure in angiosperms that consists of the basal sections of the calyx, corolla, and stamens forming a cup-shaped tube. It is also known as a floral tube, which also refers to the corolla tube and the calyx tube. It frequently contains the plant's nectaries. It is found in the majority of blooming plants, albeit the structural size and appearance vary. This distinction between hypanthium in various species is useful for identification. Some geometric forms, such as toyon, are obconic shapes, whilst others, such *as Mitella caulescens*, are saucer-shaped (Zehnder et al., 2007).

It is found in numerous families, including the Rosaceae, Grossulariaceae, and Fabaceae. In certain situations, the bloom can be so deep with such a tiny top that it appears to have an inferior ovary - the ovary lies below the other associated floral elements. The hypanthium is known by different common names in differing species. In the eucalypts, it is referred to as the gum nut; in roses, it is called the hip.

Floret

- One of the small flowers making up a composite flower head.
- One of the flowering stems making up a head of cauliflower or broccoli.

Flower

The flower is the unique reproductive structure of angiosperms. The term "flower," as it is frequently used, refers to a distinctively colored and shaped reproductive structure.

Flowers appear to have an infinite number of color, size, form, and anatomical arrangement options. They vary in size from small blossoms to massive blooms. Individual flowers in some plants, such as poppy, magnolia, tulip, and petunia, are very large and spectacular, and they are produced singly, whereas in others, such as aster, snapdragon, and lilac, the individual blooms are quite small and are borne in a distinctive cluster known as an inflorescence. All flowers, no matter how different they are, serve the same purpose: to propagate the species by generating seeds (Yamada et al., 2003).

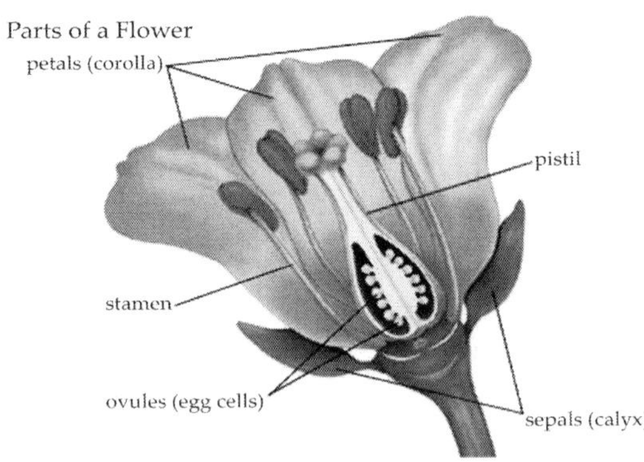

Figure 66: Parts of a flower.

Source: https://www.britannica.com/science/flower

Foliate

Predated by a number: possessing a specific number of leaflets; for example, 3-foliate, "possessing three leaflets."

- Of or relating to leaves.
- Shaped like a leaf.

Foliicolous

Foliicolous refers to the growth behavior of certain lichens, algae, and fungi that tend to develop on the leaves of vascular plants. There are about 700 species of foliicolous lichens recognized, with the vast majority of them found in the tropics.

Figure 67: Foliicolous lichens growing on a leaf on the Amazon jungle near Tena, Ecuador.

Source: https://en.wikipedia.org/wiki/Foliicolous

Follicle

In botany, a follicle is a dry unilocular fruit formed from one carpel and containing two or more seeds. Suture-induced dehiscence in order to free seeds, as seen in Consolida (some larkspurs), peony, and milkweed (Asclepias).

However, there are some difficult situations in which the term indehiscent follicle is used, such as with the genus Filipendula, which carries indehiscent fruits that could be considered halfway between a (dehiscent) follicle and an (indehiscent) achene (Yakhin et al., 2017).

In botany, a follicle is a dry unilocular fruit formed from one carpel and containing two or more seeds. Suture-induced dehiscence in order to free seeds, as seen in Consolida (some larkspurs), peony, and milkweed (Asclepias).

However, there are some difficult situations in which the term indehiscent follicle is used, such as with the genus Filipendula, which carries indehiscent fruits that could be considered halfway between a (dehiscent) follicle and an (indehiscent) achene.

Figure 68: Only some flowers in a Banksia inflorescence mature into follicles embedded in the "cone".

Source: https://en.wikipedia.org/wiki/Follicle_(fruit)

Foliole

In botany, a leaflet (also known as a foliole) is a leaf-like component of a compound leaf. Though it resembles a full leaf, a leaflet is born on a petiole or branch of the leaf rather than the main plant stem or branch, as a leaf is. Compound leaves are found in many plant families and vary greatly in shape. Palmate and pinnate leaf morphology are the two basic types of complex leaf morphology. A hemp plant, for example, has palmate compound leaves, whereas some Acacia species have pinnate leaves.

A pinnule or pinnula is the last free division (or leaflet) of a compound leaf or a pinnate subdivision of a multipinnate leaf.

Figure 69: Pinnate leaf of a legume with 10 leaflets.

Source: https://en.wikipedia.org/wiki/Leaflet_(botany)

Foliose

Foliose lichen is one of several varieties of lichens, which are complex organisms generated by the symbiotic interaction of fungi with a photosynthetic companion, usually algae. Lichen may exist in a wide range of temperatures as a result of this partnership, from freezing, dry mountains to damp, warm valleys. Lichens develop slowly, with rates ranging from 0.01mm/year to 27mm/year based on the species. Their average life expectancy is between 30 and 60 years.

The thallus, which is composed of hyphae and houses the cortex and medulla, is the main body component of lichens. The cortex contains photosynthetic cells, whilst the medulla facilitates gas exchange and makes up the majority of the lichen's thallus. There are three types of lichens: crustose, foliose, and fruticose. Foliose lichen features leafy thalli that are flattened, as well as an upper and lower cortex. Many have numerous stacked layers that help distinguish different types (Yakhin et al., 2017).

Foliose lichens adhere to surfaces by hyphae in the lower cortex, which have tiny root-like structures known as rhizines. Lichens have a vital part in the environment. They provide a food supply for many species, including deer, goats, and caribou, and they are used to create bird nests. Some species can even be employed in the production of antibiotics. They are also a good indicator of the level of pollution in the atmosphere.

Forb

Forbs (also known as herbs) are herbaceous (non-woody) broadleaf plants that are not grass-like. Rangeland forbs are typically perennial and may be poisonous, but they rarely dominate a stand unless severely overgrazed.

Cropland forbs are typically annual plants that fill seasonal gaps in high-quality feed.

Members of the Brassica genus and related genera of the Cruciferae family are important forages. Forage rape (*Brassica napus* L.), kale (*Brassica oleracea* L.), turnips (*Brassica rapa* L.), swedes (Brassica napus L.), and hybrids such as Tyfon are examples of "brassicas."

Chicory (*Cichorium intybus* L.), fodder beets (*Beta vulgaris* ssp. Vulgaris), and annual Kochia are other regularly used forbs (*Kochia scoparia* L.). *Kochia prostrada* Schrad. (perennial Kochia) is a woody shrub.

Forest

A forest is an area of terrain dominated by trees. Hundreds of definitions of forest are used around the world, with characteristics including tree density, tree height, land usage, legal standing, and biological function all taken into account. According to the Food and Agriculture Organization of the United Nations (FAO), a forest is defined as, "More than 0.5 hectares of land with trees higher than 5 meters and canopy coverage greater than 10%, or trees able to meet these criteria in situ. It does not include area principally used for agriculture or urban development." The Global Forest Resources Assessment 2020 (FRA 2020) discovered that forests covered 4.06 billion hectares using these criteria (10.0 billion acres; 40.6 million square kilometers; 15.7 million square miles), or almost 31% of the world's land area in 2020 (Wisler & Norris, 2005).

Forests are the most frequent terrestrial ecosystem on the planet, and they may be found all over the place. Only five countries contain more than half of the world's woods (Brazil, Canada, China, the Russian Federation, and the United States of America). The tropical latitudes have the highest proportion of forests (45 percent), followed by the boreal, temperate, and subtropical regions.

Forests provide 75 percent of the Earth's biosphere's gross primary output and 80 percent of plant biomass. Tropical forests are expected to produce 21.9 gigatonnes of biomass each year, whereas temperate forests will produce 8.1 gigatonnes and boreal forests will produce 2.6 gigatonnes.

Forests in various latitudes and elevations, as well as those with varying precipitation and evapotranspiration, create distinct biomes: boreal forests near the North Pole, tropical moist forests and tropical dry forests around the Equator, and temperate forests in the middle latitudes. Higher elevations support forests similar to those found at higher latitudes, and precipitation also has an impact on forest composition.

Figure 70: A conifer forest in the Swiss Alps (National Park).

Source: https://en.wikipedia.org/wiki/Forest#/media/File:Swiss_National_Park_131.JPG

Frond

The term "frond" refers to a big, split leaf. In both common English and botanical nomenclature, the leaves of ferns are called as fronds, and some botanists restrict the phrase to this category. Other botanists use the term frond to describe the broad leaves of cycads, palms (Arecaceae), and a number of other flowering plants such as mimosa and sumac. The term "fernd" normally refers to a large, compound leaf, although when applied botanically to the leaves of ferns and algae, it can also refer to smaller, undivided leaves.

Fronds use specialized terminology to define their constituents. Fronds, like other leaves, are connected to the main stem by a stalk (Weitemier et al., 2014).

In botany, this leaf stalk is known as a petiole, but in the case of fronds, it is known as a stipe, and it holds a flat blade (which may be referred to as a lamina), and the expansion of the stipe into this section is known as the rachis. Simple (undivided) blades, pinnatifid (deeply incised but not totally compound), pinnate (complex with leaflets grouped along a rachis to simulate a feather), or compounded blades (subdivided). A frond can be compounded once, twice, or three times.

Frutescent

: having or approaching the appearance or habit of a shrub: shrubby.

Fruticose

A fruticose lichen is a form of lichen fungi with a shrubby or bushy growth structure that resembles coral. It consists of a thallus and a holdfast. It is the outcome of a symbiotic relationship between cyanobacteria and two mycobionts. The vegetative structure of fruticose lichen is complicated, with an ascending, bushy, or pendulous look. While water and light are the fundamental drivers of lichen communities, vegetative propagation and filamentous growth in a fruticose lichen are typically associated with low-elevation areas. Fruticose lichens can tolerate severe desiccation. They grow slowly and are commonly found in harsh environments such as tree barks, rocky faces, and soils in the Arctic and mountains.

Figure 71: *Letharia vulpina*, a species of fruticose lichen.

Source: https://en.wikipedia.org/wiki/Fruticose_lichen

Fruit

A fruit is a developed and ripened ovary, as well as the elements of the ovary. The reproductive component of the plant flower that holds the ovules is the ovary. The ovary encloses and protects the ovules from the initial stages of floral development until they are fertilized and converted into seeds. The fruit eventually serves to disseminate the seeds or attract dispersers. Fruits can be dry or meaty, dehiscent (breaking open) or indehiscent, and single or multi-seeded (Wilkinson et al., 2012).

According to the botanical definition, many objects that are often referred to be vegetables are, in reality, fruits (for example, eggplant, green beans, okra, and, yes, tomatoes). Fruits technically contain a variety of things that we refer to as nuts (when they are enclosed by their shells, such as walnuts, sunflower seeds, peanuts, chestnuts, acorns, and even coconut). The Brazil nut, which contains a seed, is not listed.

Fruits also include some spices (allspice, vanilla bean, chili pepper, fennel seeds, black pepper, coriander), melons and squash (cucumber, pumpkin, and watermelon), the majority of berries (blueberry, cranberry, gooseberry), and grains (corn, wheat, rice), which are pretty much entirely made of seed but also contain ovary tissue.

Figure 72: the red, fleshy part of a strawberry is not ovary tissue, which makes the strawberry an accessory fruit. There are numerous ovaries (green and brown) embedded on the surface of the accessory tissue. Technically, each of these ovaries forms a dry, single seeded fruit (achene).

Source: https://www.nybg.org/blogs/science-talk/2014/08/what-is-a-fruit/

Fugacious

In botany, fugacious refers to short-lived leaves and flowers that wither and fall away from the mother plant within just a few hours or days. Some perennials are short-lived and renew once new seedlings replace the parent plant. Most perennials plants germinate, flourish, produce seedling, and die within the span of just a few weeks.

Funicle (Funiculus)

In seed plants, the ovule is the structure that gives rise to and contains the female reproductive cells

In flowering plants, the funiculus is the stalk that attaches an ovule to the placenta

Funnelform

Funneliformis is a fungus genus that belongs to the Glomeraceae family. The fungi are all arbuscular mycorrhizal (AM) fungi, which form symbiotic (mycorrhizal) relationships with plant roots (Waraich et al., 2011). In 2010, Arthur Schüßler and Christopher Walker created the genus *Funneliformis mosseae*, with *Funneliformis mosseae* as the type species (named after researcher Barbara Mosse and first identified in 1968 as a species of Endogone). The generic name refers to the funnel-shaped spore base that can be found in a variety of taxa.

G

Galbulus

Galbulus is a term used to refer to the fleshy cone. It is mostly used when relating to cone borne by cypresses and junipers. Galbulus is at times mistaken for berries. A galbulus is formed by fleshy cone scales. The cone scales accrete to form a single mass under a unified epidermis. The term galbulus was used to refer to cones produced by cypresses and is currently used to refer to those produced by junipers (Van Dijk et al., 2021).

Figure 73: Galbulus is the fleshy cone in certain plants.

Source: https://www.shutterstock.com/search/galbulus

Gall

Gall is a term used to refer to abnormal growths that occur on flowers, roots, twigs and leaves. In most cases, galls are produced as a result of stimulation or irritation of plant cells. The stimulation and irritation are as a result of insects such as mites, wasps and aphids feeding and laying eggs on the affected parts. Some galls are as a result of infections by nematodes, fungi and bacteria. This makes it difficult to differentiate between galls caused by insects and those caused by infection. However, if the insect or the eggs on the plant are observed, it becomes easy to identify the cause of the galls. Galls may take the appearance of warts, knobs, lumps or balls. The appearance is influenced by the organism causing it. Other than the appearance of the ball, they can easily be identified due to their range of color. Galls can be black, yellow, green and red. Among the factors influencing the occurrence of galls are pest populations, plant susceptibility and weather. Oaks tend to have many galls as they are more susceptible as they host a number of insects causing galls on twigs and leaves.

Figure 74: Galls are abnormal growths on plants.

Source: http://www.bio.brandeis.edu/fieldbio/Plant_Galls/samplepage.html

Gamete

Gamete is the term used to refer to sex cells of plants. Plants have both egg cells and sperms. In some cases, sperms take the form of pollen grains. These sex cells fuse resulting in the production of a zygote also called a fertilized egg. Plants are very unique as a single plant can produce both the male and female sex cells. The female gametes in plants are egg cell or ovule. They are produced by the ovary of the plant. The pollen grain enclose sperm cells. By enclosing both the male and female sex cells, they are able to retain their moisture and are protected from destruction.

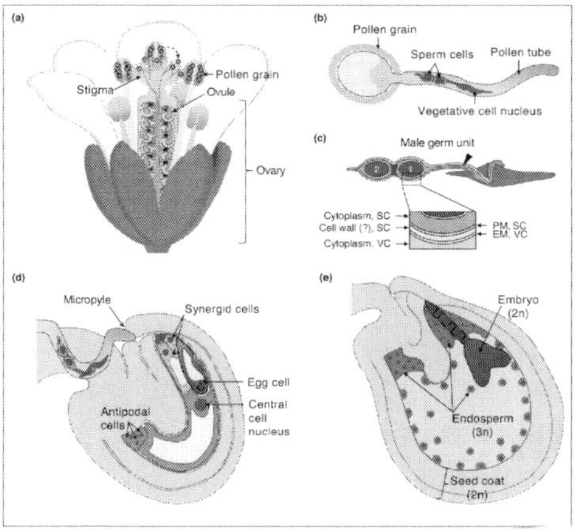

Figure 75: Sex cells in plants are called gametes.

Source: https://www.sciencedirect.com/science/article/pii/S1369526619301116

Gametophore

The gametophore in some plants is the bearer of male and female sex organs. It is also called the gametangia. The gametophore is common in fern and moss. The gametophore carries the male sex organs collectively called the antheridia and the female sex organs also called the archegonia. For some plants, both male and female sex organs are organized in the same plants. In such a case they are called monoicous. If they have only the male or the female sex organs, they are called dioicous (Van Bruggen et al., 2016). The gametophore in Bryopsida is the leafy moss plant. This is the adult form of the haploid gametophyte. Phytohormones influence the protonema which is the juvenile form to develop. In the case of the filamentous protonema, growth is through apical cell division. The division of the three faced apical cells caused the growth of the gametophore.

Gametophyte

Gametophyte is the term used to refer to a multicellular phase of life cycles in algae and plants. The gametophyte is a multicellular organism. Before it becomes a multicellular organism, it is originally a haploid spore. The haploid cells contain a single set of chromosomes. The sexual phase of algae and plants is the gametophyte. Development of the gametophyte results in the production of sex organs that produce gametes. Diploid zygotes are formed when haploid sex cells participate in fertilization. The diploid zygote contains a double set of chromosomes. The zygote undergoes cell division to form new diploid multicellular organisms. The sporophyte is the second stage in the life cycle of plants and algae. Sporophytes produce haploid spores by meiosis. After germination, they produce a new generation of gametophytes.

Gemma

Gemma is a term used to refer to a single cell that can detach itself from the parent and develop into a new individual. Gemma can also be used to refer to a modified bud of tissue or a mass of cell that can detach itself from a parent and develop into a new individual. The development is a result of asexual reproduction called fragmentation (Van Breusegem et al., 2001). Therefore plants grow due to asexual propagation. Gemma is found in mosses, liverworts, algae and fungi. Some species of butterworts and pygmy sundews are some of the flowering plants with these structures. A different kind of asexual reproduction is noted in vascular plants like turions and bulbils.

Figure 76: Gemma under suitable conditions develop to form new plants.

Source: https://en.wikipedia.org/wiki/Gemma_(botany)

Gene pool

A gene pool is the total genetic diversity found within a population or a species. A large gene pool has extensive genetic diversity and is better able to withstand the challenges posed by environmental stresses. Inbreeding contributes to the creation of a small gene pool and makes populations or species more likely to go extinct when faced with some type of stress.

Geophilous

Geophilous is a term that means earth loving and is used to plants that have gained a special adaptation that allows them to withdraw beneath the surface of the ground during diverse conditions. Some of the diverse conditions include drought and extreme heat. The adaptations may be classified as crowns, corais, bulbs and rhizomes. Rhizomes are stems found underground and they may be branched or simple. The most common rhizome is the branched rhizome. It is common because it combines vegetative reproductive reproduction alongside other advantages of rhizome habits (Van der Kooi & Ollerton, 2020).

Germination

Germination is the sprouting of spores, seeds and other reproductive bodies. Germination is normally detected after a period of dormancy. For a plant to germinate, there are some required conditions that should be met. There should be sufficient water, light and water. The seed should also experience warming and chilling. These factors also initiate germination in plants. During seed germination, the embryo absorbs water which causes the expansion and rehydration of cells. After the water absorption process, the rate of respiration in cells increases as well as other metabolic processes that reduced or were suspended during dormancy resume. All these events are linked with structural changes in the organelles found in the cells of the embryo.

Figure 77: Germination is a common event in plants.

Source: https://www.gardeningknowhow.com/ornamental/cacti-succulents/sc-gen/what-are-glochids.htm

Glaucous

Glaucous is a term used in describing a pale grey or bluish-grey appearance on some plant surfaces. It is also present as epicuticular wax coating on mature palm fruits. It is also found in grapes. Some species of cacti have the glaucous coating. The coating is hydrophobic and prevents wetting by rain.

Glochid

Glochid is also called glochidia. The term is used to refer to hair-like spines or short prickles that are found on the areoles of cacti. The short prickles are generally barbed. The glochids are easily detached from plants. They then lodge on the skin. This causes irritation of the lodge skin (Twidwell et al., 2013). Tufts have glochids found in the areoles and in some species of cactus, they nearly cover the stem. There are hundreds of glochids in each tuft. For cactus, they have tufts and conspicuous cactus sines. The cactus spines are not easily detached and are not generally barbed.

Figure 78: Glochids are commonly found in cactus.

Source: https://www.gardeningknowhow.com/ornamental/cacti-succulents/sc-gen/what-are-glochids.htm

Glumes

Glumes refers to the bract below the spikelet in the flowers of sledges or the inflorescence of grasses. In the spikelets of grasses, other than glumes there are two other types of bracts namely the plea and the lemma. Glumes in grasses make up the lowermost organs of a spikelet.

Grass

Figure 79: There are various species of grass including star grass.

Source: https://www.istockphoto.com/photos/fresh-cut-grass

Grass is a plant made up of large numbers of thin, spiky green leaves. They can cover a large area of the ground becoming groundcover.

Grassland

Grassland is an area where vegetation is dominated by grasses. In some grasslands, rush and sedge can be found with variable portions of legumes such as herbs and clover. Grasslands are common in most areas on earth with the exception of Antarctica. Various ecoregions on earth are also made of grassland. Among the largest biomes on earth is grasslands (Taiz et al., 2015). They are also said to dominate landscapes worldwide. Grasslands cover about 31-43% of the earth's land area. There are three main types of grasslands namely agricultural grasslands, semi-natural grasslands and natural grasslands.

Groundcover

Groundcover refers to plants that grow over an area of ground. Groundcover plays a crucial role in the protection of topsoil from drought and erosion. The herbaceous layer is a layer of vegetation below the shrub layer in most ecosystems. Various types of grasses form groundcovers in various areas. The term groundcover can also be used to refer to the herbaceous layer or the regenerative layer.

Figure 80: A variety of plants are planted to form ground cover.

Source: https://www.groundsguys.com/blog/2020/january/fast-growing-ground-cover/

Guard cell

Guard cells are plant cells in the epidermis of stems and leaves that are specialized to control the exchange of gases. Normally guard cells are produced in pairs. There is a gap between guard cells called the stomatal pore. When water is freely available, guard cells become turgid causing the stomatal pore to become large. However, if water is very low the guard cells become flaccid. Proper working of guard cells ensures easy diffusion of carbon dioxide into the stomata. The carbon dioxide finds its way to the mesophyll tissues allowing photosynthesis to take place.

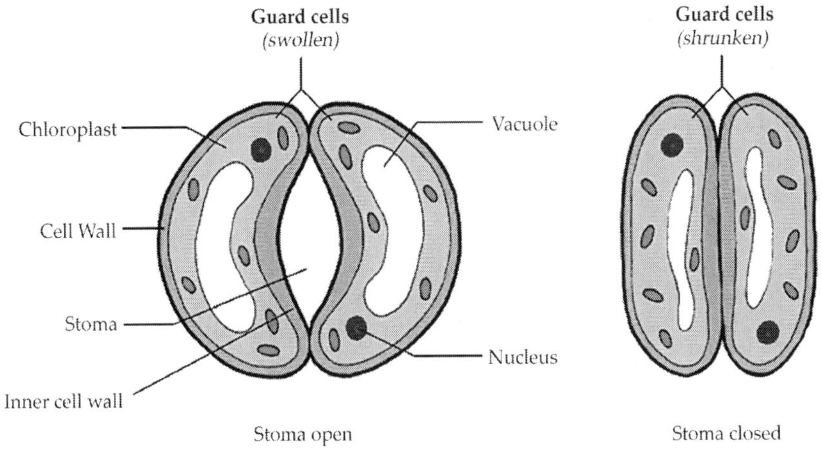

Figure 81: Guard cells allow the exchange of gases in leaves.

Source: https://www.quora.com/What-is-the-difference-between-stomata-and-guard-cells

Guttation

Guttation is the process in which drops of xylem sap are exudated on the edges of the tips of leaves of vascular plants like grasses. There are those who confuse guttation with dew. Dew is moisture that condenses on the surface of plants. Generally, guttation occurs during the night. Potassium, sugars, inorganic compounds and organic compounds make up the guttation fluid. When the fluid dries on the surface of the leaf, it forms a white crust (Szpak, 2014).

Guttulate

Guttulate is a term commonly used in botany to refer to drop –like particles or fine drops on plants or from plants.

Gymnosperm

Gymnosperm is any vascular plant whose reproduction occurs by means of exposed ovules and seeds. The opposite is the angiosperms. For these plants reproduction occurs though the seeds are enclosed by mature ovaries. Seeds of gymnosperms can also be referred to as naked seeds. These seeds are borne in cones and when they mature they become visible. Many gymnosperm plants have become extinct. For the remaining groups of plants, they have been grouped into four distinct divisions namely the Gnetophyta, Ginkgophta, Cycadophta and Coniferophyta.

Gynobasic

Gynobasic is a term commonly used to refer to the style of a flower. The style is part of the reproductive parts of a plant that arises from the base of the ovary. There are two gynobasic styles. In the first one, the style arises from the center of an apocarpous ovary. This is common in many species of Lamiaceae. The second is where the style is inserted laterally at the base of a syncarpous ovary noted in Rhabdodendraceae and Chrysobalanaceae. A common feature for the Boraginaceae is the gymnobasic style. It is important to note that the style arises from the base of the ovary.

Gynoecium

Gynoecium is a collective term used to refer to parts of a flower that produces ovules. After fertilization, the ovules germinate to form seeds and fruits. The innermost whorl of a flower is called the gynoecium. It is composed of pistils. The gynoecium is surrounded by stamens which are the pollen producing reproductive organs. Androecium is the collective name for stamens. The female part of the flower is therefore called the gynoecium (Schuurink et al., 2006). However, it produces megaspores rather than directly producing female gametes. The megaspores develop to form the female gametophyte which produces the egg cells. In botany, the term gynoecium is used to refer to a cluster of archegonia. It is also used to refer to modified stems or leaves present on a gametophyte shoot in hornworts, liverworts and mosses.

Carpellate and pistillate are flowers that bear gynoecium but no stamens while staminate is a flower that lacks a gynoecium.

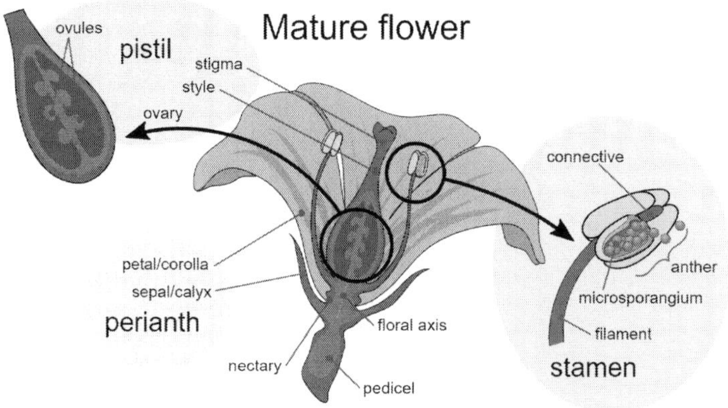

Figure 82: Female parts of a flower are called the gynoecium.

Source: https://en.wikipedia.org/wiki/Gynoecium

Gynophore

Gynophore is a part of certain flowers (stalk) that supports the gynoecium. It elevates the ovule producing part of a flower above the branching points of other floral parts. Brachychiton, *Peritoma arborea* and Telopea are some of the plant genera with flowers having gynophores.

H

Half-inferior Ovary

Half-inferior ovary is when the ovary is surrounded or embedded by the receptacle. They are also known as partially inferior, sub-inferior or half-inferior ovary. This is mostly common among flowers of the family Lythraceae. The flowers are also termed as half-epigynous or perigynous. In some cases, half-inferior ovaries are not recognized. They are grouped with other inferior or superior ovaries (Smith & Knapp, 2001).

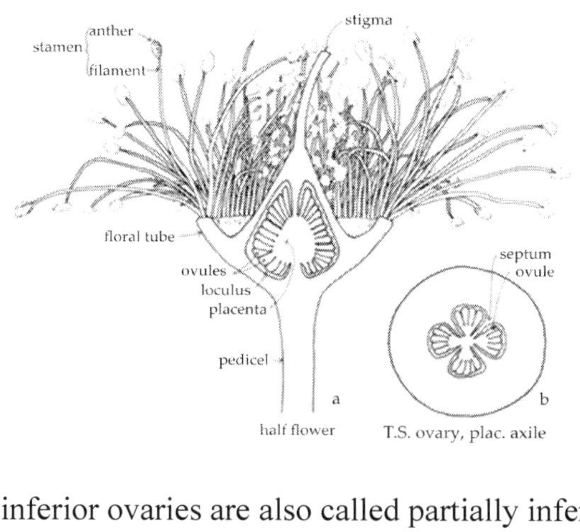

Figure 83: Half-inferior ovaries are also called partially inferior ovaries.

Source: https://science-health.csu.edu.au/herbarium/gynoecium/carpel/semi-inferior

Halophyte

Halophytes are salt –tolerant plants that grow in water or soils of high salinity. These plants come to contact with saline water by salt spray or through its roots. They are commonly found in seashores, sloughs, marshes, mangrove swamps and saline semi-deserts. Salt marsh grass is a good example of a halophyte. About 2% of all plant species are halophytes making halophytes very few plant species. The majority of plant species are the glycophtes. These plants are not salt-tolerant and are therefore easily damaged by high salinity.

Figure 84: Hydrophytes survive in saline conditions.

Source: https://dengarden.com/gardening/Hydrophytes-The-Outdoor-Garden-Water-Pond-Plants

Hand-pollination

Hand-pollination is a pollination technique when pollinating plants when the open or natural pollination is insufficient or undesirable. It is also called mechanical pollination. Hand-pollination involves the manual transfer of pollen from the stamen of a plant to the pistil of the other plant. The plant from which pollen is obtained is called the pollen parent or pollen donor and the one in which the plant receives pollen is called the seed parent (Simon & Petrášek, 2011). A small brush or a cotton swab is used during hand pollination. In other cases, hand pollination is done by removing the petals from a male flower and brushing it against the stigmas of the female flowers. In other cases, the flowers were shaken. This is done in bisexual flowers.

Haploid

Haploid is a term used to describe a cell having a single set of chromosomes. It is also used when referring to the number of chromosomes in sperm cells or egg cells collectively known as gametes. Gametes have half the number of chromosomes contained in normal diploid cells. They are also referred to as somatic cells. The parent diploid cells undergo a process called meiosis. This is a type of cell division that causes a reduction in the number of chromosomes by half. Algae and other plants have haploid portions in their life cycles.

Figure 85: Haploid cells are produced during meiosis .

Source: https://www.researchgate.net/figure/Haploid-plants-produced-by-genome-elimination-in-crosses-of-CENH3-point-mutants-by-Ler_fig7_281634855

Harmomegathy

Harmomegathy is a process in which spores and pollen grains change in shape so as to accommodate variations in the volume of the cytoplasm. The process is due to cells changing hydration.

Heathland

Heathland is a term used to refer to wide open landscapes having low-growing shrubs. Some of the shrubs include heathland grasses, heather grasses and gorse grasses. There are no herbaceous plants and few trees in heathlands. Heathlands are not natural. They are artificially created. Heathlands occur on barren lands having draining soils. The soils are sandy soils and therefore do not have enough nutrients to allow farming. Tree species such as silver birch and pine survive in these areas are properly managed. Some of the routine management practices include burning, grazing and careful cutting. To allow bare soil to restore, turf stripping is done to clear vegetation (Rusman et al., 2019).

Hemerochory

Hemerochory is the distribution of cultivated seeds, cutting and plants consciously or unconsciously by human beings into uncolonized areas through natural mechanisms of spread. These plants are well adapted to maintain themselves without any help from human beings. Hemerochory is considered to be one of the main propagation mechanisms of plants. Plants produced through Hemerochory are called hemerochoric plants affect the biodiversity of habitat by either increasing it or decreasing it.

Hemi-legume

Hemi-legumes are legume fruits with which one valve of the pod and seeds are dispersed as a unit. As with blows, the valve catches it and blows away with the seeds.

Herb

Herbs are plants used for their therapeutic properties, flavor and scent. Herbs are used to make herbal medicine that can be used as dietary supplements. Herbal medicines are sold as dried plants, extracts, teas, powders, capsules and tablets. Herbal medicines are used to improve or maintain health. Herbs have a very large use among people are they are perceived to be safe and good for people's health as they are natural. This may not be entirely true. Some herbs are dangerous to human health. They include ephedra and comfrey.

Figure 86: Herbs are plants used for their nutritional purposes.

Source: https://www.tehrantimes.com/news/465354/10-national-plans-to-promote-medicinal-plants

Herbaceous

Herbaceous refers to vascular plants having no persistent woody stems above the ground. Biennials, annuals and perennials are examples or herbaceous plants. Ferns, forbs and graminoids are examples of herbaceous plants. Forbs are characterized by broad leaves. Graminoids are vascular plants with a grass-like appearance. They include rushes, sedges and true grasses. Herbaceous plants are different from woody plants as they are low growing (Reid et al., 2009). They are also characterized by soft green stems lacking lignification while their above-ground growth is ephemeral. The opposite is herbaceous plants characterized by woody plants with stems above ground. The stems are able to survive during dormant seasons. They include woody bamboos, vines, shrubs and trees. A good example of an herbaceous plant is a banana plant as its stem does not have any true woody tissue.

Hermaphrodite

Hermaphrodite is a term used in reproductive biology to refer to organisms having both kinds of reproductive organs. Such plants are able to produce both gametes associated with the female and male sexes. This is commonly used in plants and animals. The plant, therefore, experiences a hermaphroditism condition whereby it acts as male or female.

Heteroblastic

Heteroblastic is a term used to refer to significant and abrupt changes in form and functions that occur in an animal's life span. The process affects certain characteristics including the stem structure, internode length, leaf form, leaf arrangement and leaf size. Heteroblastic change is at times confused with seasonal heterophylly. Here late and early growth in a season are marked differently. The opposite of a Heteroblastic change is the homoblastic change. The homoblastic change occurs gradually and the changes are minimal so that there is a little difference between adult and juvenile stages. Heteroblastic change is common in many different families. Convergent evolution is said to be the cause of the spread of Heteroblastic plants across species (Poff & Zimmerman, 2010).

Heterospory

Heterospory refers to the process in which spores of two different sexes and sizes are produced by sporophytes of land plants. The female spores tend to be larger than the male spores. The female spore is therefore called the megaspore while the male spore is called the microspore. The Devonian period is said to exhibit the evolvement of Heterospory from isospory. This occurred independently in several plant groups namely the progymnosperms, arborescent horsetails and clubmosses. Heterospory is considered as part of the evolution process of the timing of sex differentiation.

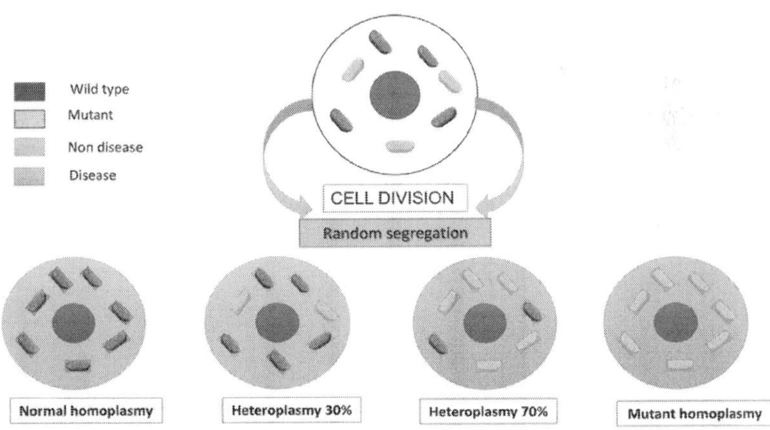

Figure 87: Heterospory involves the production of spores.

Source: https://www.researchgate.net/figure/Mitochondrial-homoplasmy-and-heteroplasmy-The-expression-of-mitochondrial-diseases-is_fig3_328639320

Heterostyly

Heterostyly is a unique from of herkogamy and polymorphism in flowers. Heterostyly occurs in heterostylous species. In the heterostylous species, there are morphological types of flowers existing in the population. The morphological flowers are called morphs. All flowers in an individual plant share the same morph. Though the morphs are similar the flower morphs have different lengths of the stamens and the pistil though the traits are not continuous (Pauchard et al., 2009). With regards to heterostyly in plants, a phenomenon called heteromorphic self-incompatibility occurs where pollen from a flower from one morph is not able to fertilize another flower in the same morph. This is because the morph phenotypes are genetically linked to genes that account for a unique system of self-incompatibility.

Husk

Husks also called hull is a term used in botany to refer to the coating or outer shell of a seed. The term husk is at times used to refer to the outer covering of the ear of corn during its growth. A husk includes the protective outer covering of vegetables, fruits and seeds. Husks can be removed after drying the harvest.

Figure 88: Rice is covered with husks during its growth.

Source: https://www.businessdailyafrica.com/bd/data-hub/turning-rice-husks-waste-into-wealth-creation-ventures-3533160

Hybrid

Hybrid is a term used to refer to plants produced as a result of cross-pollination between two different plant varieties. Therefore a hybrid is a plant that has been grown from a seed combination. Cross planting has gained popularity over the years as it allows the production of valued attributes of initial variety into the offspring. As desired attributes are cultivated in the new plant, the new plant may have improved taste, changed color, increased flowering, improved fruit production, changes in plant size and disease resistance. A large number of currently sold plants are hybrids (Prajapati & Modi, 2012). There are different kinds of hybrids of plants including inter-specific hybrids, intra-specific hybrids and inter-generic hybrids of plants. The process of creating a hybrid plant is called hybridization. In some cases, the breeder may remove stamens from plants to prevent self-pollination from occurring. Other methods are taken up to prevent cross-pollination between unwanted plants.

Figure 89: Hybrids are plants produced from seed combinations.

Source: https://www.wikihow.com/Make-Hybrid-Plants

Hybrid formula

Hybrid formula is a list containing parents involved during the creation of a plant hybrid. The hybrid formula is useful in predicting possible characteristics of the new plant hybrid. In the hybrid formula, the linking of parents is indicated by the multiplication sign. Parents are represented by their hybrid names which is the botanical name of the parent. An example of a hybrid formula is as follows:

Hybrid name * hybrid name

Hydrophily

Hydrophily is a term used to refer to an uncommon form of pollination characterized by pollen distribution through water flow more so in streams and rivers. There are two categories of hydrophilous species. The first category involves the distribution of pollen beneath the surface. It is well illustrated by seagrasses whereby pollen grains are released inside the water while the female flower is submerged in water. In the second category, pollen is distributed to the surface. This is well illustrated by the Vallisneria male flower. The flower produces pollen grain and releases it on the surface of water. The pollen is then carried away by water currents until they find their way to the female flower (Neumann et al., 2009).

Hypanthium

Hypanthium also called the floral cap is a structure in which the basal portions of the stamens, corolla and the calyx form a cup-shaped tube. The hypanthium is also called the floral tube and it refers to the calyx tube and corolla tube. The hypanthium carries nectaries of the plant. Several flowering plant species have hypanthium though differences may be noted in the appearance and the structural dimensions. The differences in hypanthium in different plant species have been used in the identification of plants. Examples of the geometric forms include the saucer shape as noted in *Mitella caulescens* and the obconic shapes in toyon.

Hysteranthous

Hysteranthous is a term used to refer to geophytes that tend to flower during dry seasons when the plant has no leaves and is dormant. It is therefore characterized by the emergence of leaves after the opening of flowers. They are the opposite of synathous plants. For these plants, flowers appear when the plant still has its leaves. Colchicum is an example of a Hysteranthous plant. Its flowers appear during fall while its leaves appear during spring.

I

Idioblast

An idioblast is a plant cell that differs from neighboring tissues. They are used for a number of things, including as storing reserves, excretory products, pigments, and minerals. They may contain, among other things, oil, latex, gum, resin, tannin, or pigments. Certain minerals contain acrid-tasting and poisonous calcium oxalate, carbonate, or silica crystals. Idoblasts are present in all plant tissues and tissue systems. Excretory idioblasts, tracheoid idioblasts, and sclerenchymatous idioblasts are the three forms of idioblasts.

Idioblasts contain biforine cells that produce crystals. The chemicals are excreted by the plant and are subsequently stored in liquid or crystalline form. They are known as druse in bundles and raphide crystals (Smith & Knapp, 2001).

The internal water pressure causes crystals or other material to be released when the end of an idioblast is ripped. Calcium oxalate idioblasts may work as a hindrance to herbivores, a means of containing or storing calcium, or a means of hardening tissue structure.

Excretory idioblasts store oils, lipids, tannins, mucilage, and minerals. They are currently being researched for their potential to store important plant therapeutic components. The selective cultivation of excretory idioblasts allows for improved collection of their stored products.

Tracheoid idioblasts are water-conducting cells that resemble tracheids in appearance. Tracheoids are elongated idioblasts with helical or reticulate auxiliary structures. They are not connected to the plant's circulatory system. Spiral cells, tracheoidal idioblasts, spirally thickened sclereids, and tracheoidioblasts are other names for tracheoid idioblasts.

Sclerenchymatous idioblasts are enlarged structural cells that provide support and rigidity to the plant. When encountered in large numbers, they are referred to as sclereids. A single sclerenchymatous idioblast is more uncommon than a group of sclereids. Their development and differentiation are unknown.

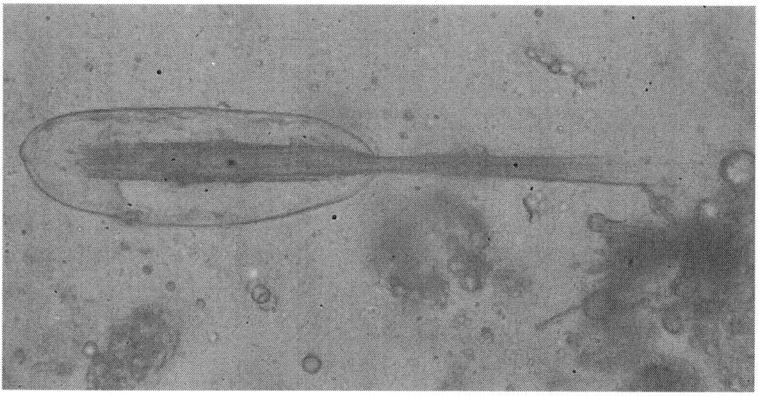

Figure 90: Calcium oxalate needles shot out from an idioblast (600x magnification).

Source: https://en.wikipedia.org/wiki/Idioblast

Imparipinnate

(A leaf shape) pinnate with a single leaflet at the apex.

The arrangement of feather-like or multi-divided characteristics growing from opposite sides of a common axis is known as a pinnation (also known as a pennation). Pinnation can be found in biological morphologies, crystals such as ice or metal crystals, and patterns of erosion or stream beds (Jeger et al., 2007).

The phrase comes from the Latin word pinna, which means “feather,” “wing,” or “fin.” “Pectination,” which is a comb-like arrangement of pieces, is a comparable idea (arising from one side of an axis only). Pinnation is used to distinguish it from “palmation,” in which the components or structures radiate out from a common point. The phrases “pinnation” and “pennation” are cognate, and while they are sometimes used differently, there is no constant difference in their meaning or usage.

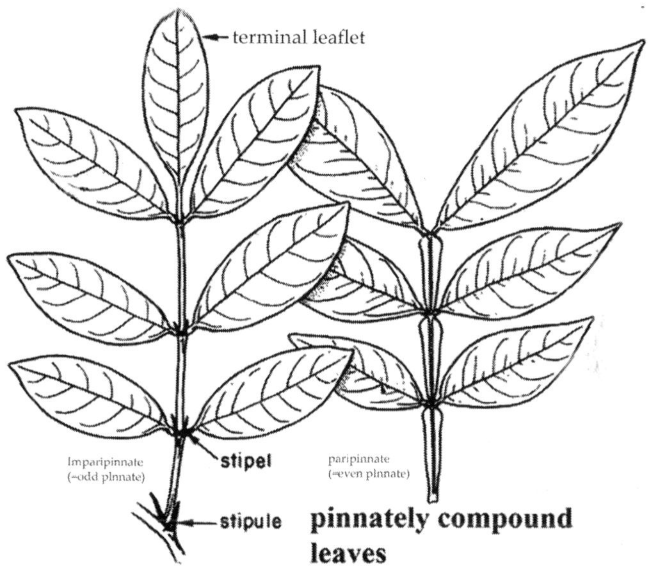

Figure 91: Imparipinnate leaf.

Source: http://sweetgum.nybg.org/science/glossary/glossary-details/?irn=1766

Inbreeding

When related individuals mate, inbreeding occurs. Self-pollination is the most extreme kind of inbreeding because the same plant serves as both the female and male parent. All plants in any closed group of plants (“closed” means no new plants from outside the group are allowed to enter and mate) will be related.

Self-pollination is another natural kind of inbreeding in plants. Inbreeding can have a considerable impact on gene expression, which can help to prevent inbreeding depression (Simon & Petrášek, 2011).

Even with flowering plants, inbreeding is often harmful. Because inbreeding increases the likelihood that defective copies of a gene will be expressed, inbred offspring have lower viability.

Inbreeding frequently produces offspring that are weaker than their parents, reducing the plant's chances of survival. Many plants rely on animals to aid in their reproduction. For example, bees transport pollen containing the male sex cell to other flowers, which fertilize the plant to produce seeds.

Incomplete Flower

A flower is said to be complete if it has sepals, petals, stamens, and pistils; a flower that is missing one or more of these components is said to be incomplete. Not all flowers have stamens and pistils. When both are represented, the flower is said to be perfect, or bisexual, irrespective of the lack of any other element that renders it insufficient). A flower that lacks stamens is said to be pistillate, or female, whereas one that lacks pistils is said to be staminate, or male (Rusman et al., 2019).

Monoecious plants (e.g., tuberous begonia, hazel, oak, corn) produce unisexual flowers of both sexes; dioecious plants (e.g., date, holly, cottonwood, willow); and polygamous plants (e.g., date, holly, cottonwood, willow) carry male, female, and bisexual flowers on the same plant.

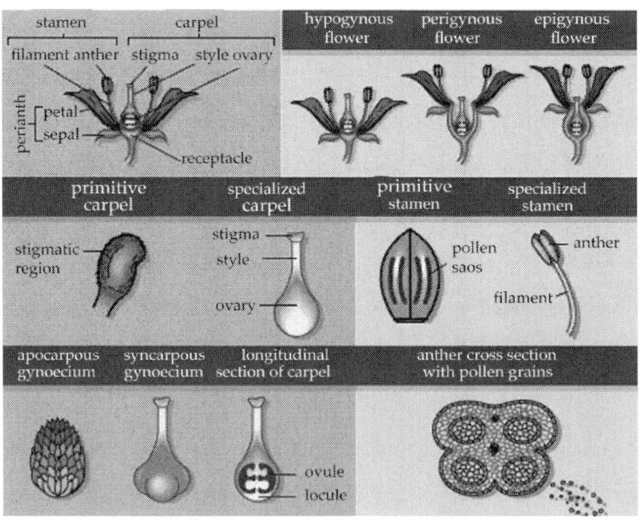

Figure 92: Floral structures characteristic of angiosperms.

Source: https://www.britannica.com/plant/angiosperm/Reproductive-structures#ref596943

A complete flower has all four components, but an imperfect flower lacks at least one of them. A bisexual (or "perfect") bloom has both stamens and carpels, whereas a unisexual (or "imperfect") flower excludes either stamens (and is referred to as carpellate) or carpels (and is referred to as staminate) (and is called staminate). Monoecious plants have both staminate and carpellate flowers on the same stem (for example, maize).

Inferior Ovary

The inferior ovary is located beneath the attachment of other floral components. A pome is a fleshy fruit that is frequently used as an example, although thorough examination of some pomes (such as Pyracantha) reveals that it is actually a half-inferior ovary. Flowers having inferior ovaries are known as epigynous flowers (Reid et al., 2009). Orchids (inferior capsule), Fuchsia (inferior berry), banana (inferior berry), Asteraceae (inferior achene-like fruit called a cypsela), and the pepo of the squash, melon, and gourd family, Cucurbitaceae, are examples of flowers with an inferior ovary.

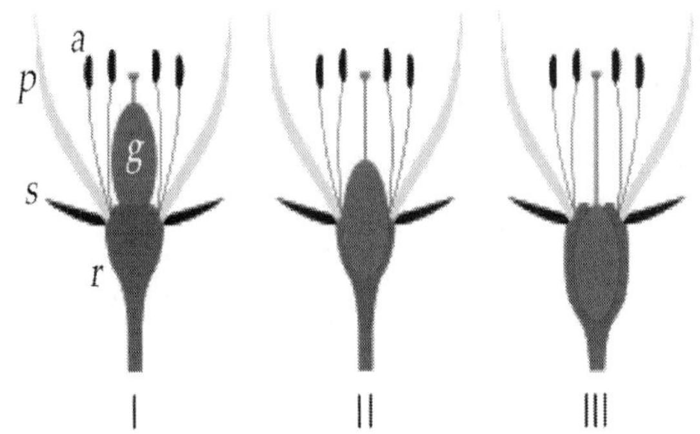

Figure 93: Ovary insertion: I superior II half-inferior III inferior. a androecium g gynoecium p petals s sepals r receptacle. The insertion point is where a, p, and s converge.

Source: https://en.wikipedia.org/wiki/Ovary_(botany)

Inflorescence

A group or cluster of flowers formed on a stem made up of a primary branch or a complicated arrangement of branches is referred to as an inflorescence.

According to morphology, flowers are produced in the modified part of the stalk of seed plants. The length and nature of the internodes, as well as the phyllotaxis, can all be changed, as can the proportions, compressions, swellings, adnations, connations, and reduction of the main and secondary axes. Inflorescences are the reproductive portion of a plant that produces a cluster of flowers in a certain pattern (Poff & Zimmerman, 2010).

The stem that supports the entire flower is known as a peduncle. The rachis is the primary axis (incorrectly called the main stem) above the peduncle that contains the blooms or smaller branches. The stalk of each flower in an inflorescence is known as a pedicel.

Figure 94: *Aloe hereroensis*, showing inflorescence with branched peduncle.

Source: https://en.wikipedia.org/wiki/Inflorescence

A lone flower is a flower that is not a component of an inflorescence, and its stem is referred to as a peduncle. A floret is any flower in an inflorescence, particularly if the separate flowers are little and produced in a tight cluster, as in a pseudanthium. The fruiting stage of an inflorescence is known as an infructescence. Simple (single) or complex (many) inflorescences (panicle). There are several types of rachis: solitary, composite, umbel, spike, and raceme (Pauchard et al., 2009).

Infructescence

Infructescence (fruiting head) is the collection of fruits produced by an inflorescence's ovaries. It normally keeps the inflorescence's size and structure.

In certain circumstances, infructescences resemble simple fruits in appearance. These are known as many fruits. The infructescence of Ananas, for example, is created by the union of the berries with the receptacle tissues and bracts.

An ear is the mature infructescence of a grain, such as wheat or maize. Ficus infructescence is known as a syconium.

Figure 95: Ananas (pineapple) infructescence.

Source: https://en.wikipedia.org/wiki/Infructescence

Infundibular (Infundibuliform)

Infundibular means "funnel-shaped" in plant morphology, for example, referring to the corolla of some flowers.

Insectivorous

Insectivorous plants, often known as carnivorous plants, get some or all of their sustenance from catching and eating animals or protozoans, most typically insects and other arthropods. Photosynthesis, on the other hand, provides energy to carnivorous plants. Carnivorous plants have evolved to flourish in situations with thin soil or low nutrient levels, notably nitrogen, such as acidic bogs. Insectivorous Plants, the first well-known dissertation on carnivorous plants, was written by Charles Darwin in 1875. Except for Antarctica, carnivorous plants can be found on all continents and several Pacific islands (Prajapati & Modi, 2012).

True carnivory is thought to have evolved nine times separately in five different groups of flowering plants, with several genera displaying it. At least 583 species in this category entice, trap, and kill prey, utilizing the available nutrients as a result. The quantity of species has risen by roughly three every year since the year 2000. Furthermore, around 300 protocarnivorous plant species from various genera exhibit some but not all of these features. According to a 2020 assessment, nearly one-quarter of all species are endangered with extinction as a result of human actions.

Figure 96: An upper pitcher of *Nepenthes lowii*, a tropical pitcher plant that supplements its carnivorous diet with tree shrew droppings.

Source: https://en.wikipedia.org/wiki/Carnivorous_plant#/media/File:Murud_N._lowii_20.jpg

Intercalary Meristem

Plants have a form of tissue called the meristem. It is made up of undifferentiated cells (meristematic cells) that can divide. Cells in the meristem have the ability to develop into all of the other tissues and organs found in plants. These cells continue to divide until they become differentiated, at which point they lose the ability to divide.

The intercalary meristem is a type of meristem found in certain plants, most notably grasses. These cells, like apical and lateral meristems, can divide and generate new cells. They differ in that they are found among mature tissue sections, such as at the base of grass leaves, which are positioned on mature stem tissue. Intercalary meristems often exist for just a short time before developing into mature tissues (Neumann et al., 2009).

Intercalary meristems are present at nodes on the branch where leaves have grown and are essential for grass shoot and leaf elongation. Intercalary meristems are internode sites in which ground meristem cell division occurs for a longer period of time than in other sections of the internode. Rosette plants lack intercalary meristems.

Intercalary (also known as basal) meristems are found in monocot (especially grass) stems at the base of nodes and leaf blades in angiosperms. Intercalary growth is also seen in horsetails and Welwitschia. Many monocots use intercalary meristems for rapid growth and renewal because they are capable of cell division. Intercalary meristems at bamboo nodes allow for quick stem elongation, whereas those at the base of most grass leaf blades provide for rapid regrowth of damaged leaves. This leaf regeneration in grasses originated in response to grazing animal damage.

Interjugary Glands

In pinnate leaves, glands occurring along the leaf rachis between the pinnae (occurring below the single, and often slightly larger, gland at or just below the insertion of the pinnae).

Internode

A vascular plant's stem is one of two primary structural axes, the other being the root. It provides support for leaves, flowers, and fruits, carries water and dissolved chemicals between roots and shoots in the xylem and phloem, stores nutrients, and generates new living tissue (Katova, 2007).

Normally, the stem is separated into nodes and internodes:

The nodes contain one or more leaves, as well as buds that can develop into branches (containing leaves, pine cones, or inflorescences (flowers)). The nodes may also develop adventitious roots.

Internodes are the nodes that connect one node to another.

Internode: the space between two subsequent nodes. It has the ability to extend, depending on the species, either from the base or from the extremity.

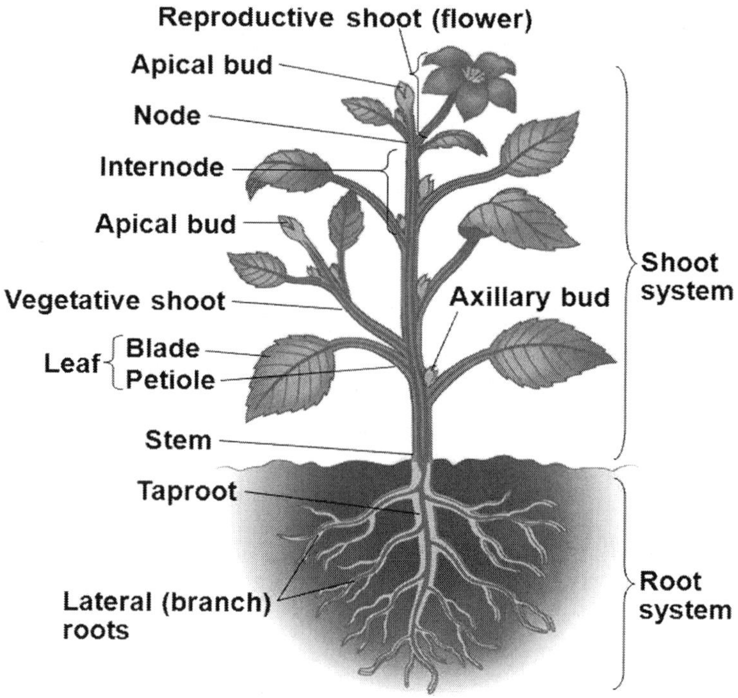

Figure 97: Plant structure.

Source: https://www.quora.com/What-are-the-functions-of-the-nodes-and-internodes-in-plants

Interpetiolar

A stipule is a plant protrusion that grows on both sides (or sometimes only one side) at the base of a leafstalk (the petiole). Stipules are regarded part of the anatomy of a normal blooming plant's leaf, yet in many species they may be inconspicuous — or even completely absent, in which case the leaf is referred to as exstipulate. (In some ancient botanical writings, the term "stipule" was used to refer to any little leaves or leaf-parts, particularly prophylls.) Linnaeus created the term stipule from the Latin stipula, which means straw or stem.

Free lateral stipules, adnate stipules, interpetiolar stipules, intrapetiolar stipules, ochreate stipules, foliaceous stipules, tendrillar stipules, and spiny stipules are all types of stipules (Nikunj et al., 2014).

A stipule is "interpetiolar" if it is positioned between the petioles rather than attached to the petioles, and one stipule from each leaf is usually fused together, giving the impression that there is just one stipule between each leaf.

Intramarginal

Inside but close to the margin, for example a vein in a leaf.

Figure 98: The intramarginal veins near the margins of this leaf are outlined in white.

Source: https://en.wikipedia.org/wiki/Glossary_of_botanical_terms

Intrastaminal

Inside but close to the margin, for example, a vein in a leaf.

Involucre

In botany, a bract is a customized or specialized leaf that is attached to a reproductive component such as a flower, inflorescence axis, or cone scale. Bracts are typically (but not always) separate from the leaves of the foliage. They could be smaller, larger, a different color, shape, or texture, or all of the above. They frequently differ from other floral features such as the petals or sepals. A plant with bracts is called bracteate or bracteolate, whereas one without bracts is called ebracteate or ebracteolate (Novello et al., 2020).

An involucre is a whorl of bracts found at the foot of an inflorescence. Involucres are found below the inflorescences of many Apiaceae, Asteraceae, Dipsacaceae, and Polygonaceae. In this scenario, each flower in an inflorescence may have its own whorl of bracts, known as an involucel.

A plant with bracts is referred to as bracteate or bracteolate, while one without bracts is referred to as ebracteate or ebracteolate.

Isotomic

Having branches of equal diameter.

J

Jugary Gland

A gland found on the rachis of a pinnate or bipinnate leaf on a jugum, as in some Acacia species, or the junction or attachment of pairs of pinnae or pinnules (Klassen, 2001).

Juvenile Leaves

Leaves formed on a young plant, typically differing from the adult leaves in form.

K

K, K+, K-

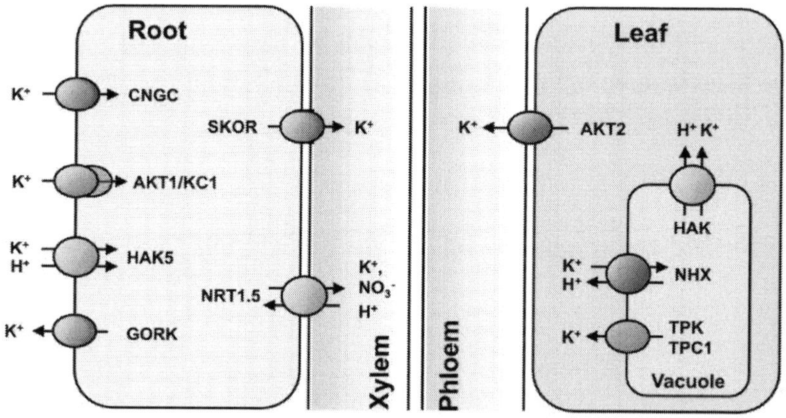

Figure 99: Potassium (K+) isn't integrated into natural matter yet stays as a solvent compound in the cell sap consisting of up to 10 percent of the dry natural content.

Source: https://www.frontiersin.org/articles/10.3389/fpls.2019.00281/full

For lichenology studies, "K" is the truncation for the result of a test where a 10% mix of potassium hydroxide (KOH —thus "K") is set on lichen tissues. Color variation is distinguished by "K-" for none, plus K+ for a yellow to crimson or purple tone. K is bigger than K+. Since when a single electron is eliminated from K particle, it has just 18 electrons. Be that as it may, the nucleus contains 19 protons. So the protons will draw in the electrons with a more prominent affinity. Consequently, K+ particle shrivels. K is bigger than K+ (Nikunj et al., 2014).

The range of the neutral K molecule is bigger than the cation K+ particle. This is on the grounds that the size of the particle relies upon the distance of the peripheral electrons from the core. These electrons have "greater energy levels".

The range of the potassium particle relies upon the area of the single electron, available on the fourth energy level that is the most, from the nuclei.

The cation K+ develops when K loses this peripheral single electron to achieve a unique "electronic setup". The K+ particle has its furthest electrons on the third energy level presently and its size is more modest than that of the K particle. Thus, K is bigger than K+.

Keel

A conspicuous longitudinal edge like the boat keel, for example, the design of the corolla developed by the combination of the lower axis of the pair of abaxial front petals of blossoms in the Fabaceae. The lower section of a papilionaceous blossom (starting at a pea or bean plant) comprises a pair of fused petals, lying between the wings, and encases the pistil and stamens. Keel is the bottom petal of the pea-like bloom, is short and circular.— herbal science: the lower part of a papilionaceous blossom (a pea or bean plant) that comprises of two intertwined petals, lying between the wings, and encases the pistil and stamens.

Kernel

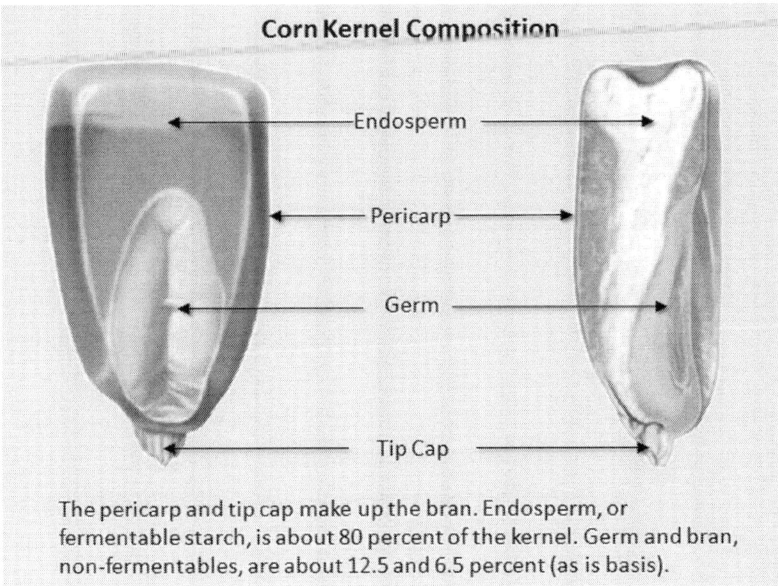

Figure 100: The Corn Kernel consists of five unique parts which are the endosperm, pericarp, germ and tip cap.

Source: https://www.pinterest.com/pin/438678819928335974/

The edible piece of a nut that is inside the shell, or the entire seed of a grain plant: 1.The fundamental piece of a seed; all that is inside the seed dividers; the edible substance contained in the shell of a nut; thus, anything contained in a shell, husk, or integument; as, the part of a nut. 2. A single seed or grain; as, a part of corn. 3. A little mass around which another matter is cemented; a core; a solidification or hard knot in the tissue (Novello et al., 2020).

Kettle trap

Another word for the kettle-based pitchers of any of the predatory pitcher plants, in which they trap their prey. Kettle trap blossoms have a deceptive fertilization structure, including the deploying of general remuneration showing mixtures to trap both male and female, transcendently nematocerous Diptera, which looks for nectar or other sugar-rich plant exudates.

Elaborate kettle trap blossoms to briefly confine pollinators developed autonomously in a few angiosperm genealogies. Research on types of Aristolochia and Ceropegia as of late enlightened how these specific kettle trap blossoms draw specific pollinators through compound duplicity.

Kidney seeds

Figure 101: As the name suggests these beans have a distinct kidney shape that can't be mistaken.

Source: https://www.neseed.com/shop/vegetable-seeds/beans/shell-dry/dark-red-kidney/

Term for a reniform item like a bean or leaf; all the more officially, oblately cordate, or bow formed with the corners rounded. Kidney beans constitute a form of the common bean (*Phaseolus vulgaris*), a legume local to Central America and Mexico. Common bean is a significant food source and wellspring of protein all through the world. Utilized in a variety of customary dishes, kidney beans are generally eaten all around cooked (Mercuri et al., 2002).

Poorly or inappropriately cooked kidney beans are poisonous, yet wholesome beans can be a sound part of a meal routine. They come in an assortment of colors and forms, including white, purple, cream, dark, red, striped, mottled, and spotted.

Kingdom

The highest classification of the taxonomic system, over that of division (phylum). The Plant Kingdom incorporates vascular plants, bryophytes and green growth and is otherwise called the clade Viridiplantae. Plantae is a plant kingdom containing all plants found on the earth. They are multicellular eukaryotes. Distinctively, they contain a tough structure that encompasses the cell film known as the cell wall (Matyjaszczyk & Śmiechowska, 2019). Plants additionally have a green-shaded color known as chlorophyll that is used for photosynthesis. The plant kingdom is ordered into five subgroups as indicated below: Bryophyta. Pteridophyta. Thallophyta. Gymnosperms. Angiosperms.

Klausenfrucht

Klausenfrucht (German) is a unique kind of fruit product in Lamiaceae and Boraginaceae. A dried, dehiscent fruit shaped from a prevalent ovary with axil or basal placentation, and with a follower calyx, from more than one carpel and ordinarily falling to pieces into 1-cultivated units by isolating every carpel by bogus septa. One unit is a half carpel, generally, there are four units, seeds. Eremocarp, schizocarp, mericarp or nutlets.

This organic product is a unique type of decay fruit, a break fruit product, which occurs for the most part in the mint family (Lamiaceae) and scavenger leaf family (Boraginaceae). It deteriorates to maturity by parting genuine and bogus allotments into generally forlorn fractional fruits.

Knee

Figure 102: A root that projects over the ground, ordinarily happening in species that grow in waterlogged grounds. Breathing root or pneumatophore, whereby the horizontal root develops a circle that arises at low tide.

Source: https://www.healthbenefitstimes.com/glossary/knee-root/

A sudden curve in a root or stem, usually at a node; a pneumatophore or cypress knee, is a kind of twist or handle in the base of certain plants, particularly conifers, for example, a part of the Taxodioideae, that shows as a projection of the root over the ground level or mud level. The knee roots, similar to the pneumatophores, emerge from even roots and develop upwards until they enter the soil surface. Their uncovered parts bear lenticels. Large numbers of the knee roots retain roots underneath the soil surface and re-enter the soil (Mandel, 2019).

Knee roots are created by Bruguiera species.

Knee roots arise as a root circle from the underground root structure and permit the trading of gases in oxygen-deprived residue. Every underground evenly developing root fosters a few knee roots at customary spans.

Knee root foundations of a grown-up *Bruguiera gymnorhiza* for instance stretch out in a range of roughly 10 meters around the trunk compartment and can come to a height of up to 60cm.

The knee roots of the distinctive Bruguiera species contrast in size, shape and frequency plus can change based upon the area and development conditions.

Most knee roots have various lenticels with the capacity to exchange gas.

L

Labellum

One of three to five petals which is (generally) not quite the same as the others, for example in Orchidaceae, Cannaceae, Stylidiaceae, and Zingiberaceae. In plant science, the labellum (or lip) is the piece of the blossom of an orchid or Canna, or other less-known genera that serves to draw in bugs, which fertilize the bloom, and serve as an arrival stage for them. Labellum (plural: labella) is the Latin minor of labrum, which means lip.

The floral families with one stamen and a differed number of petal-like staminodes show an extraordinary scale of flower structure. The most prominent piece of the bloom in the family Zingiberaceae is generally called the labellum; this design, nonetheless, doesn't comprise of similar flower parts as the labellum in the family Lowiaceae and in orchids, in which it is a petal. The labellum consists of a few fused stamens, and in numerous genera, there is a couple of more modest however obvious petal-like staminodes, each addressing one stamen (in the family Zingiber they are attached to the labellum); the outcome is a bloom of orchidlike form, yet the parts found in orchid and ginger blossoms are unique, and without a doubt, the two have matured along very unique developmental lines. In the families Marantaceae and Cannaceae, the single stamen is just half useful, the other half is pretty much petal-like. In the two families, the staminodes are of differing size and shape, so their status isn't settled. In the family Costaceae, the enormous labellum addresses a relation of every one of the five nonfunctional stamens, and sometimes it very well may be five-lobed (Lindström et al., 2013).

Labiate

Figure 103: The mint group's (Labiatae) "lipped" petals. The circular corolla is separated into two undetachable parts, typically with two flaps in the upper lip and triple projections in the lower lip.

Source: https://www.healthbenefitstimes.com/glossary/labiate/

Lipped; where a corolla is isolated into two sections, known as the upper and lower lip, the two looking like an open mouth having lips.

- Having the appendage of a rounded corolla or calyx separated into two uneven parts, one projecting above the other same as the sage, catnip and snapdragon.
- Belonging to a natural plant order (Labiatae), of which the sage, mint, and catnip are models. They are for the most part sweet-smelling spices

Lacerate

A laceration is a cut that is caused by the tearing of delicate plant tissue. This kind of wound is frequently sporadic and rough. The laceration is frequently contaminated with microscopic organisms and germs from whatever object caused the cut. Contrarily, a stab is typically brought about by a sharp item like a nail, organism teeth, or tack. This kind of wound can seem to shut for the day. Punctures are additionally susceptible to disease and ought to be treated accordingly (Luan et al., 2020).

Laciniate

Of flaps – with edges sporadically separated into sharply isolated, thin, pointed sections; of edges – profoundly partitioned into pointed fragments in an unpredictable way. 1. laciniate - having edges sporadically and delicately cut; "a laciniate leaf" bordered. Course- of the edge of a leaf shape; having the edge cut or bordered or scalloped.

Lamella

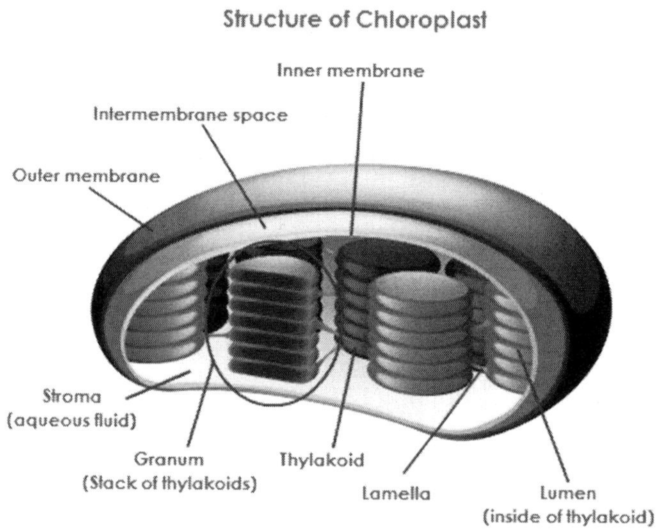

Figure 104: Lamellae have a layer inside the film within the stroma in grana stacks and a complete piece of the photosynthetic responses. The other section, the light-independent responses occur within the liquid stroma.

Source: https://socratic.org/questions/in-plants-what-is-the-stroma

A slim, plate-like coating. (Plural lamellae; adjective lamellate – made out of a group of layers). A sheet-like film found inside a chloroplast of the autotrophic cell. They function as a sort of wall whereby chloroplasts can be fixed inside, achieving the best lightweight conceivable. The central lamella is a layer that concretes together the essential cell walls of two connecting plant cells. The central lamella is comprised of calcium and magnesium pectates (Lambers et al., 2008). In a full-grown plant cell it is the furthest layer of cell divider. In plants, the gelatins form a unitary and consistent layer between contiguous cells. The lamellipodia and lamella comprise of plate-like augmentations of the cell that assume essential parts in both cell motility and movement, and mechanosensing. These units form and operate under particular conditions.

The lamella or lamellum confines behind the lamellipodium and is normally the widest structure in motile cells (ordinarily 10-15 μm wide). Comprising principally of consolidated straight actin packages, the actin fiber structure of the lamella is steadier and less unique than that of lamellipodia and may likewise oppose pressure. Lamellae include more grounded and fully grown attachment sites, and furthermore contain myosin II, an engine protein that is fundamental in cell motility.

Lamina

The sharp edge of a leaf or the extended upper layer of the bract, sepal, or petal. In Hymenophyllaceae, the lamina (leaf edge) is generally just a single cell layer dense among the veins. The sori are situated at the apex or along the edges of the leaf sections and are encased in a cup-molded to narrow cone-shaped protective covering of tissue (indusium) opening outward (Klassen, 2001). In leaf morphology, a wide extended cutting edge (the lamina), is joined to the plant stem by a stalklike petiole. In angiosperms leaves ordinarily have a couple of constructions known as stipules, which are situated on each side of the leaf base and may resemble spines, glands, leaf-like structures and scales. The blade covers a major photosynthetic plant surface and seems green and leveled in a plane opposite to the stem.

Lanceolate

Figure 105: Lancelot is a medium-sized bush with dull green foliage and a thick circular growth pattern. Lancelot roses are famous for flower bundles and other decorative designs. The plant is dry season resistant yet may require water in the initial planting summer and in arid conditions.

Source: http://www.namethatplant.net/gallery_glossary_5.shtml?term=lanceolate

Oval and tapering to a sharp tip. The (plan) state of a basic or unifoliolate leaf, or for separated (compound) leaves on this alludes to the state of the leaflets. Remember that the depiction of leaf shape is, best case scenario, inexact, considering that there exists a full range of variety between each typical shape (for example, widely ovate, rough orbicular). Further difficulties emerge from a variety between taxon, inside a singular specimen, and on account of compound leaves, between individual pamphlets. When leaves are toothed or lobed, leaf/handout shape is approximated by a line drawn with regards to the apices of every tooth/flap (Katova, 2007).

Latex

Milky sap. A smooth liquid present in many plants, for example, poppies and spurges, which radiates when the plant is sliced and coagulates when exposed to the air. The rubber tree latex is the main origin of natural rubber. Artificial latex is commonly used to make coatings, paints and so forth.

Latex is a compound derived from certain types of trees, which is utilized to make items such as rubber and paste. Organic latex is formed from the *Hevea brasilienesis* rubber tree and is the safe liquid contained underneath the bark. It is gathered by cutting a tiny piece of bark from the tree and permitting the latex to ooze into a vessel over a few hours.

Natural latex with a high solid content is likewise utilized for making molds for concrete, wax, casting mortar, low temperature metals, and polyester. Natural latex can contract around the item to be imitated, with the goal that the slightest detail will be duplicated in the cast. Latex is in any case, being utilized to assist with balancing out desert soils to make them applicable for agrarian uses (Jeger et al., 2007).

Natural latex is created from the *Hevea brasilienesis* rubber tree, which is the liquid contained underneath the bark. It is an overcast white fluid, comparative in appearance to cow milk.

Hevea trees grow fully at 5 to 7 years old and may be tapped for as long as 30 years. Rubber produce rates at around a ton for every ace of land (2.5 tons per ha) on the bigger plantations, however, yields four-fold the amount as hypothetically conceivable. Trees frequently are replenished for a period following heavy tapping.

Natural latex was once industrially produced in the Amazon in large amounts. As of late, the creation of natural latex has moved to Malaysia, Indonesia, and other Far Eastern regions. Over 90% of the global production of natural rubber emanates from Asia, with more than half of that figure beginning in these nations. Other top Asian producers incorporate Sri Lanka, Thailand, and India. The Philippines and China have significantly expanded their rubber production also.

Lax

Loose and thickly packed (typically applied to the blossoms in an inflorescence). The plant chemical auxin and its delivery system are known to assume a big part in characterizing the embryonic node and ensuing improvement of the body plan. Albeit the job of PIN auxin efflux carriers has been allotted during undeveloped shoot and root development, the task of the auxin transporters AUX1 and LIKE-AUX1 (LAX) proteins isn't grounded (Hirschberg, 2001). Synthetic and hereditary instruments on Brassica napus microspore-inferred undeveloped organisms and Arabidopsis thaliana zygotic incipient organisms, and show that AUX1, LAX1 and LAX2 are needed for both shoot and root post-development, working together with PIN efflux transporters. Besides, there's a positive-criticism circle between MONOPTEROS (ARF5)- subordinate auxin signaling and transportation.

Lignify

To become woody

- To create woody tissue because of incrustation of lignin in auxiliary development.
- To become tough or fixed, similar to something constructed of wood.
- To transform into wood; to become ligneous.

Lipstick plant

Figure 106: Local to Malaysia and Indonesia, the Lipstick plant is a tropical plant with conspicuous foliage and attractive red blossoms. It is additionally grown as a houseplant to upgrade the appearance of your gallery or lounge area.

Source: https://balconygardenweb.com/lipstick-plant-care-and-growing-guide/

Aeschynanthus Radicans, generally known as the lipstick plant, is a common evergreen tropical houseplant. It derives its name from the dynamic red and rounded blossoms that show up over a burgundy bud. These blossoms grow in groups and, alongside the waxy, lustrous, green foliage, they have a falling, plant like structure. This gives them an optimal decision for use in hanging crates or tall holders (Hadwiger, 2013). An epiphytic species, these plants are local to the tropical districts of Southeast Asia. Here they can be observed developing from tree limbs and in breaks in rocks. If they get sufficient warmth, moisture, and filtered daylight, you'll enjoy a productive showcase of blooming through a significant part of the year. By and large, however, lipstick plants blossom most richly in the mid-year and fall. One thing actually significant is that the lipstick plant isn't one to choose to add a fragrant smell to your home. While the blossoms don't actually have a strong aroma, it isn't especially wonderful!

Indeed, even with its great, tropical appearance, the lipstick plant is for the most part viewed as a generally simple to-cultivate houseplant. Everything revolves around getting it right as far as lighting and dampness levels. Lipstick plants like natural however sifted light. An excess of direct daylight can cause leaf burn, and too little will bring about a negative situation of blossoming and leaf drop.

In their local clammy and tropical areas, these plants develop as epiphytes frequently growing onto branches, rock hole and other different plants. Care is needed to guarantee they don't experience the ill effects of root decay when planted up. Pruned lipstick plant will gain from being filled in a medium that is all around circulated air through, equitably soggy, and light.

In spite of the fact that lipstick plants prefer moderate dampness, especially during their most productive developing period, overwatering and soaked conditions can prompt root decay, leaf drop, and parasitic issues. Moderate watering is ideal. In a perfect world, you need to abstain from permitting the preparation medium to dry out totally and offer water when the top two or three inches are not soggy (Guo & Liu, 2007).

Lanate

Shrouded in or made out of woolly hairs. Plant hairs (scientifically called trichomes) might be seen on leaves or stems. They are extended epidermal cells; on a stem or leaf that might be living or dead, deciduous or continual; on a root it is fleeting and restricted to the absorption area, known as a "root hair". Leaves and stems without hairs are known as glabrous. Different plant parts comprise of an epidermis. It is the furthest layer of cells of youthful plant parts; on roots, a few cells separate into root hairs, on stems and leaves into hairs of different forms (pubescence) and on leaves, berries and stems into stomata. Pubescent plant parts are shrouded to put it plainly, fine, delicate hairs. Tomentose parts have a furry covering of short, firmly tangled hairs.

Plants perform gas exchange on small pores known as stomata. Carbon dioxide from the air enters the stomata and oxygen created by photosynthesis diffuses out of the stomata. Water atoms likewise escape through the stomata, particularly in sweltering, dry climate. Moisture loss through the stomata is called transpiration. Assuming that the plant loses a lot of water it will wither and in the end perish. To adapt to this difficulty, plants have advanced protective cells on each side of the stoma (Franklin et al., 2013). When the gatekeeper cells are completely bloated or extended, they have a stretch opening (stoma) between them. The walls adjoining the stoma are extremely slim and adaptable, while the external dividers on the opposite sides of the stoma are a lot thicker and more inflexible. This differential thickening makes for a flexible opening when the gatekeeper cells are swelled by inner water pressure (called turgor pressure). When the watchman cells lose water on a hot day, they become collapsed and push together, subsequently stopping the stoma. This strategy keeps the plants from losing inordinate moisture through transpiration.

Lanceolate

Longer than expansive, barely ovate, widest in the lower half and tightening to the tip, similar to a spear or lance head

- Having the overall state of a spear; significantly longer than wide, with the vastest part lower than the center and a sharp summit.
- Of a class of knapped stone places, made without a stem, shoulders, scores, or different elements that guide in connection to a shaft.

Lanceolate means a leaf, sepal, petal, or other level design that is more extensive at the base than at the midpoint, tightens toward the peak, and has a length-to-width proportion of 3:1 or more. It has its vastest point beneath the center. The various types of lanceolate leaves are determined by the length to wide proportion as follows:

Ovate = length to width proportion from 1.5:1 to below 2:1; broadly ovate = length to width proportion from 1.2:1 to under 1.5:1, highly ovate = length to width proportion of 1:1 or less; barely ovate = length to width proportion of 2:1 to under 3:1; and lanceolate = length to width proportion of more than 3:1.

Lateral

Connected to the side of an organ, for example, leaves or branches on a stem. A plant sidelong organ, characterized as either a leaf or a leaf-like organ of the shoot or bloom, emerging from a group of starter cells inside the flanks of the shoot apical meristem (SAM) or flower meristem. For instance, the tobacco leaf is developed from a set of ~100 primary cells in every one of the three histogenic layers of the SAM. Parallel organ primary cells follow fates which are totally different from the more focal SAM or adjoining cells that will develop tissues of the shoot or flower pivot (Foxcroft et al., 2017). They transform their series of extension and division and elaborate new structures of development—away from the SAM and along the side. As a parallel organ develops, cell division stops and cells separate. The developed leaf ordinarily comprises of a few million cells and ~20 diverse cell types. In any case, it shows a trademark shape, size and example of tissues. For instance, the Antirrhinum leaf, which is typical of the species, shows a designed plan of tissues that is especially obvious along the dorsiventral (adaxial–abaxial) organ node and in the conveyance of veins. Patterning is likewise found in the nonrandom appropriation of epidermal cell forms, such as trichomes and stomatal protection cells.

Two differentiating mechanisms are generally conjured to clarify such patterned development and cell personality: either that it is a characteristic capacity of primary cells, where case daughter cells should gain information about their characters from their parents; or that cells gauge their position and react to it, which infers that fate determination includes cell–cell signaling.

Latex

Figure 107: Latex is generally an opaque white sap, for example, in poppy, rubber tree, milkweed, and fig tree yet some plant latex is liquid and additionally has unique colors even in species firmly related.

Source: https://www.treehugger.com/where-does-latex-come-from-3269790

A smooth milky liquid that oozes from such plants like spurges, dandelions, and figs. Latex is commonly a clear white sap, for example, in milkweed (Asclepias syriaca, Apocynaceae), poppy (Papaver somniferum, Papaveraceae), fig tree (Ficus carica, Moraceae) and rubber tree (Hevea brasiliensis, Euphorbiaceae), yet some plant latex is direct and has particular tones even in related species.

Plant latex and different exudates are saps that are retrieved from the areas of plant damage caused either physically or by foraging insects. While most plant species ooze latex or exudates, and whereas the cautious jobs of plant latex against herbivorous bugs have for quite some time been recommended by a few investigations, the point by point roles and elements of different latex fixings, proteins and synthetic substances, for anti-herbivorous plants safety has not been very much archived in spite of the wide availability of latex in the plant realm. As of late, notwithstanding, considerable headway has been made. A few latex proteins, including cysteine proteases and chitin-related proteins, have been figured to contain significant protective abilities against herbivorous insects (Fahad et al., 2017).

For the mulberry (Morus spp.)– silkworm (Bombyx mori) connection, an ancient and notable model arrangement of plant–bug cooperation, plant latex and its fixings – sugar-emulate alkaloids and protective protein MLX56 – are found to assume key parts. Complex atomic interactions between Apocynaceae species and its symbiotic herbivores, in which cardenolides and protective proteins in latex assume key roles, are turning out to be increasingly obvious. Arising perceptions recommended that plant latex, similar to an animal toxin, is a depository of valuable protective proteins and synthetic compounds that has developed through interspecific cooperations. Then again, expert herbivores created modern transformations, either atomic, physiological, or social, against latex-borne guards. The presence of different variations in expert herbivores itself is proof that latex and its compounds work as protection basically against generalists.

Laticiferous

Laticifers are greatly specialized cells creating a cylindrical structure all through the plant body, happening in phylogenetically inconsequential groups. Laticifers create and store latex which is delivered upon bursting of laticifers. Plants ooze latex sap in light of actual harm. These comprise of dense walled, enormously extended and much stretched pipes containing a smooth or yellowish shaded juice known as latex. They contain various nuclei which lie inserted in the partly coating layer of cellular material. They are sporadically circulated in the frame of parenchymatous cells.

Laticifers are profoundly specialized cells causing a cylinder-like organization structure all through the plant body, happening in phylogenetically inconsequential groups. Laticifers produce and store latex that is delivered upon a burst of laticifers. Compounds predeveloped in the latex, like peptidases, chitinases, and chitin-restricting proteins, assume significant defensive parts against microorganisms and even bugs. Latex peptidases arise as defensive particles in more than one way. Synergism and mutual work including artificially diversified latex ingredients allow the plant to be protected from natural enemies.

Latrorse

A sort of anther dehiscence where the anthers open along the side toward adjoining anthers. cf. extrorse, introrse. Connecting with anthers that release or split toward the edge, toward different anthers, and not toward or away from the focal node around which a bloom is organized.

The situation with Ceratophyllum is upheld by a shift from latrorse to introrse or extrorse anthers (in both taxa, the thecae are moved toward one surface, however regardless of whether this is abaxial or adaxial is obscure). When the Pennipollis plant and Appomattoxia are both connected with Ceratophyllum, the three taxa can be joined by introrse or extrorse (rather than latrorse) anthers and halfway infratectal structure.

Laurel forests

Lauroid or laurophyll leaves are defined by a substantial mix of wax, giving them a tight, pointed oval shape with an apical mucro or "dribble tip", which allows the leaves to shed water notwithstanding the moisture, permitting respiration.

Laurel and laurophyll woodlands have an inconsistent dispersion in warm temperature locations, frequently involving geographical refugia where the humidity from the sea gathers with the goal that it falls as downpour or haze and soils have high humidity levels. They have a gentle environment, only occasionally presented to flames or ices and are found in moderately corrosive soils (Elmer & White, 2018). Basic efficiency is high, yet can be restricted by gentle summer dry spells. The casings are evergreen, controlled by species with polished or rugged leaves, and with modest tree variety. Bugs are the main herbivores, yet birds and bats are the dominating seed-dispersers and pollinators. Meanwhile, decomposers like invertebrates and microorganisms on the forest floors are essential to nutrient recycling.

Some laurel woodlands are a form of the cloud forest. These cloud forests are common on hill slopes where the dense humidity from the ocean or sea is hastened as warm soggy air masses brushing off the sea are constrained upwards by the landscape, which cools the air mass to the dew point. The dampness in the air consolidates as downpour or haze, making an environment defined by cool, soggy conditions in the air and soil. The subsequent environment is wet and gentle, with the yearly wavering of the temperature directed by the closeness of the sea. Laurel woodslands are portrayed by evergreen and hardwood trees, growing up to 40 m (130 ft.) in tallness. Tree woods, laurisilva, and laurissilva all establish networks that look like the bay laurel.

A few plant categories have a place with the genuine tree family, Lauraceae, however many have comparable foliage to the Lauraceae because of concurrent advancement. As in some other rainforest, plants of the tree woods should adjust to high precipitation and moistness. The trees have adjusted because of these environmental drivers by creating comparable structures, leaves that repulse water. Laurophyll or lauroid leaves are described by a liberal layer of wax, showing up, and a restricted, pointed oval shape with an apical mucro or "dribble tip", which allows the passes on to shed water regardless of the moistness, permitting breath. The systematic names Laurina, Laurifolia, Laurophylla, Lauriformis, and Lauroides are regularly used to name types of other plant families that look like the Lauraceae. The term Lucidophyll, alludes to the sparkly surface of the leaves, and was proposed in 1969 by Tatuo Kira.

The word "lauroid" is likewise applied to climbing plants, for example, ivies, whose waxy leaves fairly take after those of the Lauraceae. Mature shrub woodlands regularly have a thick tree shade and low light levels at the backwoods floor. Some timberlands are described by an overstory of rising trees (De Waele & Elsen, 2007). Shrub woodlands are regularly multi-species, and diverse in both the quantity of species and the genera and families represented without any solid environmental specific pressure, the quantity of species sharing the arboreal layer is high, albeit not arriving at the variety of tropical timberlands; almost 100 tree species have been detected in the laurisilva rainforest of Misiones (Argentina), around 20 in the Canary Islands. This species variety diverges from other mild woods varieties, which ordinarily have a shelter taken by one or a couple of animal groups. Species variety by and large increases towards the tropics. Thereby, the laurel forest is a momentary form between mild woodlands and tropical rainforests.

Lax

Loose, not dense. Of heaps of hyphae in lichens – not solid and not adglutinate. There is increasing proof that plant tissues specific for auxin transport correspond with adjacent maxima in the auxin level. This outcome is hard to accommodate with the conventional model of auxin delivery, which depends on significant degrees of auxin efflux delivery expression. Since moving cells sustain significant amounts of auxin efflux, one anticipates that a reduction of auxin is comparative to encompassing tissues.

Leaf

Figure 108: Leaves are quite possibly the primary organs that plants have. Photosynthesis, the process by which plants produce food utilizing light, carbon dioxide (CO_2), and water, happens in leaves. The makeup of leaves is intended for photosynthesis.

Source: https://cid-inc.com/blog/leaf-area-how-why-measuring-leaf-area-is-vital-to-plant-research/

An outgrowth of a stem, typically level and green; its primary function is food production by photosynthesis. Leaf, in natural science, any naturally leveled green outgrowth from the stem of a vascular plant. As the primary source of photosynthesis, leaves fabricate nourishment for plants, which thusly eventually feed and support all land organisms. Naturally, leaves are a basic piece of the stem structure (Chen & Yamaguchi, 2002). They are appended by a persistent vascular structure to the remainder of the plant so that the free exchange of supplements, water, and ultimate results of photosynthesis (oxygen and sugars specifically) can be conveyed to its different parts. Leaves are started in the apical bud (developing tip of a stem) alongside the tissues of the actual stem. Certain organs that are cursorily altogether different from the typical green leaf are framed in a similar way and are really changed leaves; among these are the sharp spines of prickly plants, the needles of pines and different conifers, and the asparagus tail or a lily bulb.

The primary role of a leaf is to deliver nourishment for the plant by photosynthesis. Chlorophyll, the substance that establishes a leaf's trademark green tone, consumes light energy. The inner structure of the leaf is protected by the leaf epidermis, which is persistent with the stem epidermis. The focal leaf, or mesophyll, comprises of delicate walled, unspecialized cells of the form known as parenchyma. As much as one-fifth of the mesophyll is made out of chlorophyll-containing chloroplasts, which assimilate daylight and, related to specific chemicals, utilize the brilliant energy in deteriorating water into its components, hydrogen and oxygen (Clemens et al., 2013). The oxygen freed from green leaves replaces the oxygen eliminated from the air by plant and organism breath and respiration. The hydrogen acquired from water is joined with carbon dioxide in the enzymatic cycles of photosynthesis to create the sugars that are the premise of both plant and organism life. Oxygen is passed into the environment through stomata—pores in the leaf surface.

Lecanorine

Of lichens, containing apothecia with rims of tissue like the tissue of the thallus, as in the Lecanora genre. A lichen contains lecanorine fruiting body parts assuming they are formed similar to a plate with a ring surrounding them, and that ring consists of tissue like the primary non-fruiting lichen body part. The name emanates from the lichen genre Lecanora, whose members contain such apothecia. In case a lichen has lecanorine apothecia, the actual lichen is occasionally depicted as being lecanorine.

Legume

Figure 109: Legumes consist of lentils, chicken peas, beans, and peanuts. They develop in a pod which later rapture to release the seeds.

Source: https://www.healthline.com/nutrition/legumes-good-or-bad

A legume is a crop in the family Leguminosae or the natural product or seed of such a plant. When utilized as a dry grain, the seed is additionally called a pulse. Legumes are developed agronomically, basically for human utilization, for animal feed and silage, and as soil-improving green excrement. Notable legumes incorporate tamarind, horse feed, clover, chickpeas, peanuts, lentils, lupins, beans, soybeans, peas, mesquite, and carob. Legumes produce an organically special kind of fruit product – a basic dry natural product that forms from a basic carpel and normally dehisces (opens across a crease) on different sides.

Legumes are eminent in that the vast majority of them have mutual nitrogen-fixing microbes in structures known as root knobs. Thus, they assume a key part in crop rotation.

- A fruit product of the family Fabaceae, shaped from one carpel or dehiscent along the two sides, or indehiscent.
- A plant species in the Fabaceae family.
- A crop within the family Fabaceae.

Lemma

The bottom of 2 bracts encasing a grass blossom. Lemma is a phytomorphological word alluding to a piece of the spikelet. It is the bottommost of a pair of bracts encasing the grass floret. It frequently bears a long fiber known as an awn, and might be comparable in structure to the glumes—chaffy bracts at the foundation of every spikelet. Sterile lemmas are formed between the rich floret and the simple glumes. The sterile lemma has not been hereditarily portrayed and is frequently confused with a glume.

Lenticel

Generally lenticular (lens formed) permeable tissue in the bark with huge intercellular spaces that permits direct trade of gases between the inward tissues and environment through the bark. The lenticels available on the epidermis of various plant organs (stem, fruits, petiole) comprised of parenchymatous cells are pores that consistently stay open, rather than stomata, which manage their degree of opening (Craigie, 2011). Lenticels are noticeable on organic product surfaces, like mango, apple, and avocado. Lenticels license the trading of gases between the climate and the inward tissue sections of the organs (stems and a few fruits). They allow for the entry of oxygen and all the while the emission of carbon dioxide and moisture. In apple fruit products, lenticels represent roughly 21% of respiration.

The lenticels which are developed in specific spaces of the phellogen (the lenticular phellogen) are natural and organically fundamental elements of the suberised periderm that permit for gas trade between the climate and the underlying tissues. The lenticular phellogen activity proceeds without fail, corresponding with that of the encompassing phellogen. The lenticels consequently take up a radially adjusted cylindrical structure, extending from an external point to the phellogen.

M

Maculate

Figure 110: Maculate plants are plants having spots.

Source: https://nurserybuy.com/product/drimiopsis-maculata-african-false-hosta/?_route_=product%2Fdrimiopsis-maculata-african-false-hosta%2F&v=c86ee0d9d7ed

Maculate is a term used to refer to plants known to have spots. Most of these plants are grown for decorative purposes. Examples of plants with decorative foliage are *Aucuba* and *Pulmonaria*. The *Dactylorhiza maculata* is another example of a maculate plant. The plant is characterized by a moorland spotted orchid. The plant is of the family Orchidaceae and is also an herbaceous plant. The plant is also characterized by an erect, glabrous and cylindrical stem. Its surface is streaked with the leaves having dark-ellipsoid-shaped spots on the surface.

Male flower

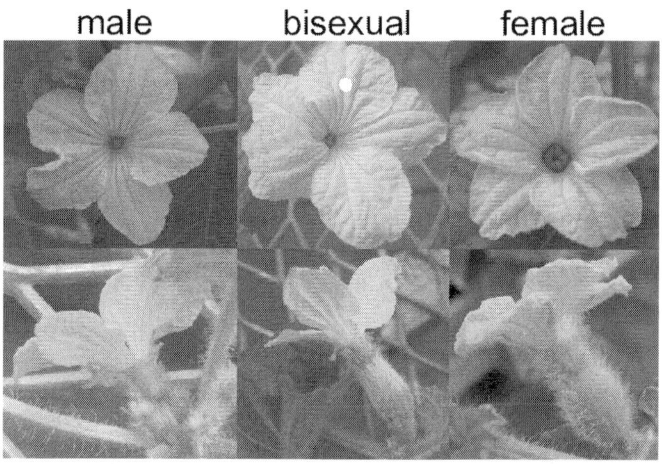

Figure 111: Male flowers bear male organs in the plant.

Source: https://www.researchgate.net/figure/Top-and-side-view-of-male-cv-Hales-Best-Jumbo-bisexual-cv-Hales-Best-Jumbo_fig1_237102133

Male flowers are flowers that only bear male parts. The male parts are collectively known as the stamen. The stamen is composed of filaments and anthers. The male flowers produce the pollen grains used to fertilize eggs cells. Pollen grains are carried by the anthers and are usually yellow in color. The filament carries and holds up the anthers. Like the female flowers, male flowers also have sepals and petals. For insect-pollinated flowers, the petals are colored to enable them to attract pollinators. The sepals are very useful in protecting the developing bud. They are green in color and are found at the base of the flowers.

Mallee

Mallee also known as *Jasminum sambac* is a plant native to tropical Asia. It is a species of jasmine. The plant is a vine or a small shrub. It grows to a height of 0.5 to 3m. Mallee is extensively cultivated. It is considered a sweet fragrant flower. The flowers are used in making jasmine tea and perfumes.

Mangrove

Mangrove is a tree or shrub that grows in brackish water or coastal saline. The mangrove also refers to a tropical coastal vegetation type composed of such species. A characteristic of these plants is that they are tolerant of salt. For this reason, they are also known as halophytes. The plants are suitably adapted allowing them to live in harsh coastal conditions. Mangrove plants are made up of complex salt filtration systems (Chandler & Sanchez, 2012). The root system is suitably adapted to enable them to cope with saltwater wave action and immersion. Mangrove plants are well adapted to the low oxygen conditions in their growing areas. For this reason, these plants are able to adapt to the intertidal zone. The mangrove forest is also called the mangrove biome which is a distinct saline shrub-land or woodland composed of a depositional coastal environment. Mangrove plants are able to survive in a given range of saline conditions.

Figure 112: A mangrove forest.

Source: https://www.istockphoto.com/search/2/image?phrase=mangrove+forest

Marsh

Marsh is a term used to refer to an area covered by water. It is also called a wetland. There are a number of plant species that grow in marsh. Plants form the structural foundation for wetlands. Certain species of plants are used to characterize marsh lands. There are different species of plants depending on the marsh area. They include the salt-to-brackish marsh plants, freshwater marsh plants and peatland plants.

Mast

Mast is a fruit produced by shrubs and forest trees. They include nuts and acorns. The term was initially used to mean the nuts of forest trees that have accumulated on the ground. Some of the nuts were historically used in fattening domestic pigs and in feeding domestic animals. The various species of trees in forests in the tropics of Southern Asia are said to mast at irregular periods ranging between 2-12 years. For woody species of plants, mast is considered the reproductive part of the plant and also the vegetative part of the plant (Bourgaud et al., 2001).

Figure 113: Nuts and acorns are examples or mast.

Source: https://www.iplantz.com/plant/1273/polyalthia-longifolia/

Medulla

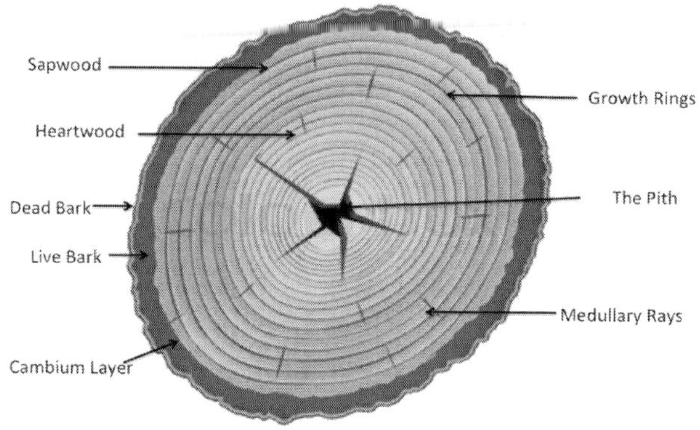

Figure 114: The medulla is also called pith.

Source: https://treegrowthstructure.weebly.com/cross-section-of-a-tree.html

Medulla also called pith is a tissue found in the stems of vascular plants. The medulla is made up of soft and spongy parenchyma cells. These cells are very useful in the storage of starch. For eudicotyledons, the medulla is found at the center of the stem. This is different in monocotyledons whereby the medulla extends from the center into the flowering stems and roots. A ring of xylem encircles the medulla. The xylem is in turn encircled by a ring of phloem. The new medulla growth is characterized by white or pale color. As it matures, the tissue gets a darkened deep brown color. For trees, the medulla is found in young growing plants. However, in the trunk, the medulla is replaced by the xylem.

Megasporangium

Megasporangium is a term used in botany to refer to the structure of a plant body having female reproductive organs. The megasporangium is at times called the ovule. The megasporangium is composed of the integument and nucellus. The megasporangium can be formed in different shapes and sizes (Arruda et al., 2015). The form of the megasporangium changes as the flower develops. As the megasporangium matures, it can be grouped into five groups namely the hemianatropous, amphitropous, camphylotropous, anatropous and atropous. The megasporangium after maturity forms the seed coat and is at times involved in the supply of nutrition on the plant.

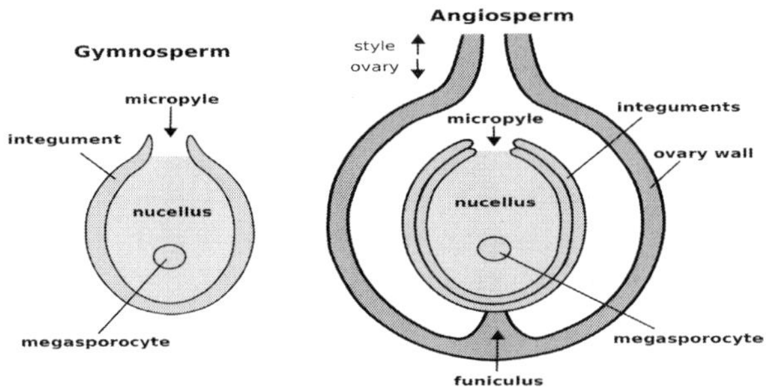

Figure 115: Megasporangium is the female reproductive organ in plants.

Source: https://plantlet.org/the-megasporangium-and-the-female-gametophyte/

Megaspore

Megaspores are a type of spore produced by heterosporous plants. They are also called macrospores normally there are two main types of spores produced by plants namely the microspores and megaspores. The megaspores are large spores. They develop to form the female gametophytes which produce egg cells. The microspore produces the male gametophytes which produce sperms that fertilize the egg cells. Examples of plants that produce megaspores include the seed plants, quillworts, spikemosses and water ferns.

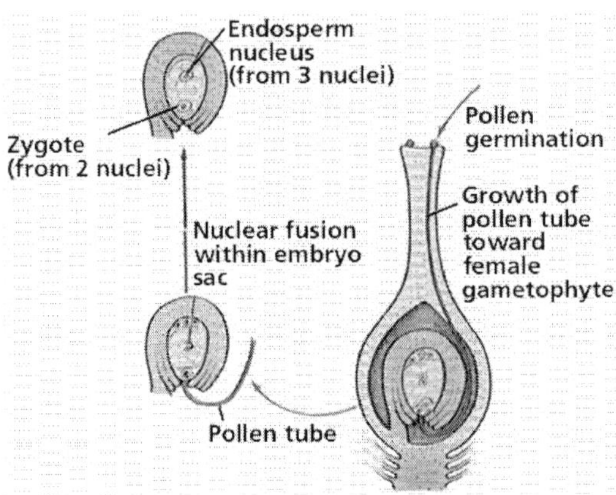

Figure 116: Megaspores are involved in the production of eggs cells.

Source: http://www2.estrellamountain.edu/faculty/farabee/biobk/biobookflowers.html

Megasporophyll

Megasporophylls are ovules produced by leaf homologs of female plants. Cycads are plants generally known to produce megasporophylls. However, the megasporophylls of cycads do not form true cones. The megasporophylls are arranged into two or three whorls at the stem apex. The stem is able to resume vegetative growth as the megasporophylls are interposed between whorls of the cataphylls and foliar leaves (Zehnder et al., 2007). Normally the arrangement is two or three whorls of leaves. The whorls of the cataphylls are then followed by megasporophylls. However, there are variations in the sequences. The megasporophylls are composed of an expanded terminal blade, a fertile portion that bears two to six ovules and a stalk.

Megastrobilus

Megastrobilus is a term used to refer to a female cone. It is also called ovulate cone or a seed cone. Megastrobilus contain ovules that become seeds after fertilization. The structure of megastrobilus varies between conifer families. The structure of megastrobilus of a plant is used to identify the plant. The term megastrobilus is from the term strobilus that refers to an organ on plants under the division pinophyta and contain reproductive structures. Megastrobilus are characterized by two types of scales namely the seed scales and bract scales.

Figure 117: The megastrobilus is also called the seed cone.

Source: https://depositphotos.com/stock-photos/megastrobilus.html

Mericarp

Mericarp is a carpel with one seed. It also refers to one pair split apart when it matures.

Meristem

Meristems are tissues found in plants. They are made up of undifferentiated cells also called meristematic cells. These meristem cells are able to undergo cell division. The meristem cells are able to develop into other organs and tissues found in plants. Meristem cells undergo cell division until they lose their differentiation ability. Once a plant cell is differentiated, it cannot produce cells of a different type. They also cannot divide into different types. Meristematic cells are incompletely differentiated or undifferentiated. They are also capable and totipotent of continued cell division. Division of meristematic cells results in the production of new cells involved in the differentiation and expansion of tissues. It also facilitates the initiation of new organs that give the plant its structure. Several meristem cells make up the meristem. Meristem cells may not have small vacuoles while the protoplasm completely fills the cell.

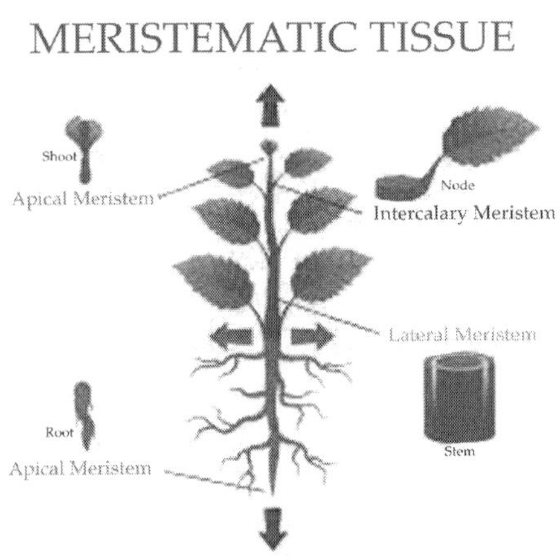

Figure 118: Meristems differentiate to form certain organs in plants.

Source: https://www.istockphoto.com/search/2/image?phrase=meristem

Mesocarp

Mesocarps are parts of a plant made up of parenchyma cells and of a diameter of 50- 300 mm. mesocarps are normally enclosed within a rigid cell wall. The parenchyma cells in mesocarps contain inorganic compounds, water-soluble substances, polyphenols, dissolved sugars and oil droplets. Mesocarps are considered the edible part of the fruit. It contains vitamins, amino acids and fatty acids.

Mesomorphic

Mesomorphic is a term used to refer to leaves that are specially adapted to conditions of relative humidity and abundant water. They are different from xeromorphic which are plants that are specially adapted to dry conditions while hydromorphic plants are those that were adapted to aquatic situations (Yamada et al., 2003).

Mesophyll

Mesophyll is a term used to refer to tissues in a plant. There are different types of mesophyll in plants each suitably adapted to perform a specific function in plants. For instance, the palisade mesophyll layer is involved in the absorption of light energy as the tissues are suitably adapted to absorb light. These tissues are found in leaves. The palisade mesophyll layer is made of cells that are towards the upper surface of the leaf, are column-shaped, closely arranged and are packed with chloroplasts. The spongy mesophyll tissues are involved in gaseous exchange. These tissues are loosely packed to allow efficient gaseous exchange. A thin layer of water covers the spongy mesophyll. The thin layer of water enables gases to dissolve in it as the gases move in and out of the cell. During the day, photosynthesis takes place. The leaf is suitably modified so that carbon dioxide diffuses into the spongy mesophyll cells as oxygen diffuses out of the cells. For gases to find their way in and out of the cells, the stomata are actively involved. They open and close allowing the entry and release of gases.

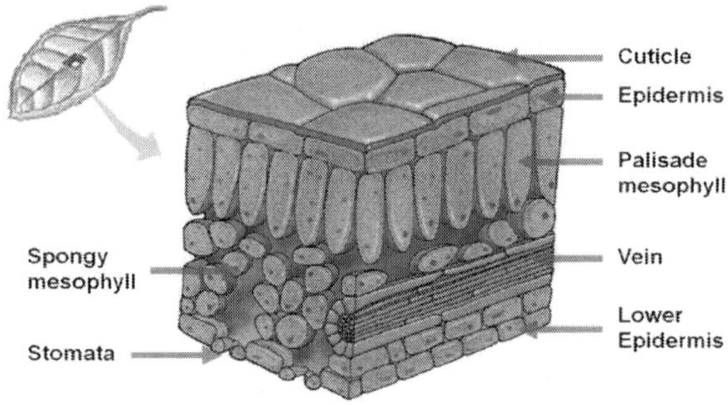

Figure 119: The spongy mesophyll.

Source: https://www.pathwayz.org/Tree/Plain/CROSS+SECTION+OF+A+LEAF+%5BBASIC%5D

Mesophyte

A mesophtye is a terrestrial plant that is not adapted to survive in particularly wet or particularly dry environments. A rural temperate meadow is an example of a mesophytic habitat. Some of the plants that grow in such areas include *Rosa multiflora*, oxeye daisy and goldenrod. Mesophytes are plants that prefer air and soil of moderate humidity. They tend to avoid soils with abundant salts and standing water. Terrestrial plants make up the largest ecological group. These plants survive in moderately humid and hot climatic regions.

Micropyle

Micropyles are pores in the membrane covering of an ovum. The micropyle allows the entry of sperms. It is therefore an important element as it allows fertilization take place.

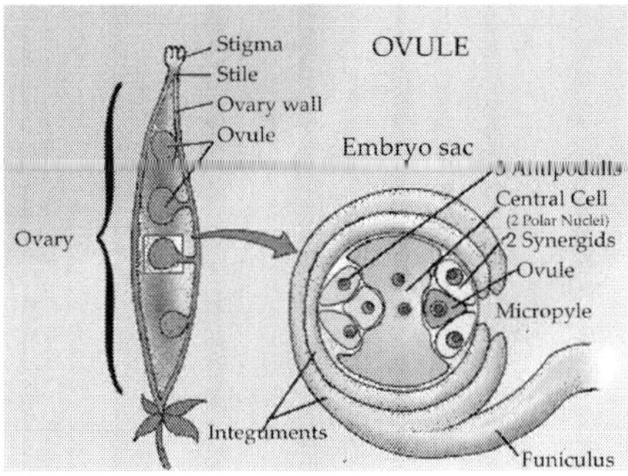

Figure 120: The micropyle is an entry way for sperms allowing fertilization

Source: https://www.cactus-art.biz/note-book/Dictionary/Dictionary_M/dictionary_micropyle.htm

Microsporangium

Microsporangium is a sporangium that produces microspores during germination that give rise to male gametophytes. Microsporangium is common in all vascular plants having heterosporic life cycles. They include the aquatic fern genus Azolla, spike mosses and seed plants. Microsporangium in angiosperm and gymnosperm anthers are involved in the production of microsporocytes. The microsporocytes through meiosis produce four microspores (Yakhin et al., 2017). The anthers of angiosperm and microsporangia of gymnosperms produce the microsporocytes. Microsporangium is a diploid microspore mother cell that produces four haploid microspores through meiosis. They then develop to form pollen grains. The microspores undergo division through mitosis to produce simple gametophytes.

Microspore

Microspore is a land plant spore that develop into male gametophytes. They are different from megaspores which develop into female gametophytes. The male gametophytes produced by microspores give rise to sperm cells that are involved in the fertilization of an egg cell resulting in the formation of a zygote. During the Devonian period, plants with heterosporous life cycles used microspores and megaspores that arose independently in several plants. Meiosis is the process by which microspores are produced from diploid microsporocytes. Microspores are therefore haploid.

Microsporophyll

The microsporophyll is a leaf-like structure bearing the microsporangia. They include those of the strobili of lycophytes. Microsporophylls are also found in male cones of conifers. In flowering plants, the stamens are highly modified microsporophylls.

Microstrobilus

Microstrobilus is a term used in botany to refer to the smaller kind of cone produced by gymnosperm. The microstrobilus is the male cone and produces pollen. Compared to the megastrobilus, the microstrobilus is very small. It is considered an aggregation of many small structures.

Midrib

The midrib is the central, thick, linear structure of a leaf. It runs along the length of the plant lamina or thallus. Midribs are normally found in true leaves. In true leaves the midrib vein runs from the leaf base to the apex. They are leaf-like structures found in seaweeds and mosses. The midrib functions as a translocative vessel. The midrib supports the whole leaf. It allows the transportation of water and minerals from the soil to the leaves. The other veins are all interconnected to the midrib. The midrib is stronger than other veins found in the leaf.

Figure 121: The midrib is the main vein that runs from the base apex to the tip of the leaf.

Source: https://www.sciencephoto.com/media/30444/view/leaf-mid-rib-and-veins

Midvein

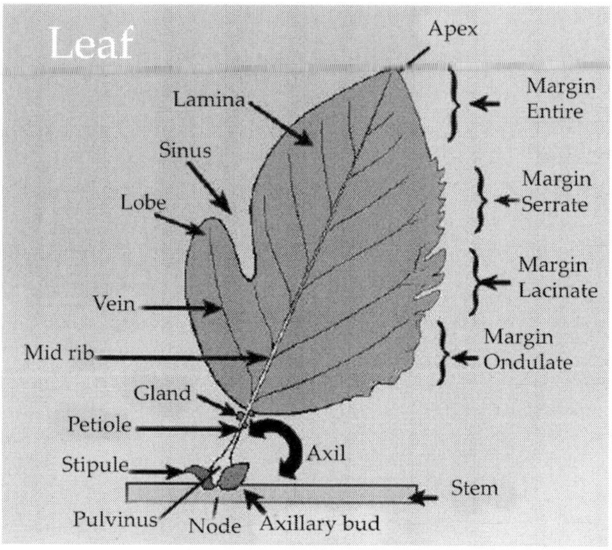

Figure 122: The midvein is also called the stem of the leaf.

Source: https://www.cactus-art.biz/note-book/Dictionary/Dictionary_M/dictionary_midvein_midrib.htm

The midvein also called the midrib is the central vein or main vein of the leaf. The midvein arises from the lateral veins and the secondary veins. Midveins can be prominently depressed or raised. The midvein is continuous with the petiole. It is also raised above the leaf blade. For compound leaves the midvein extends from the first sets of leaflets to the end of the leaf. The midvein is actively involved in the transportation of water and mineral salts from the stem to other veins in the leaf. It acts as a link between the veins and the stem. The midvein is relatively stronger than other veins found in the leaf. The function of the midvein is to support the entire leaf (Yakhin et al., 2017).

Monad

Monad is a term used to refer to pollen released by angiosperms. Monads are released at the anther dehiscence. They can be aggregated pollen that have developed over the years during the evolution of flowering plants. Pollen can aggregate as polyads, tetrads and dyads. For angiosperms, the most common form of aggregation is the tetrads. The development of aggregated monads in angiosperms is a highly conserved process in angiosperms and it is also well characterized.

Monocarpic

Monocarpic is a term that refers to plants that flower, set seeds and die afterward. The terms semelparous and hapaxanth have the same meaning as monocarpic. Polycarpic is the antonym of monocarpic and it refers to plants that flower and sets seeds several times in their lifetime (Wisler & Norris, 2005). Plietesials are plants that flower before dying. Monocarpic plants can live a number of years before they flower. Flowering in plants is an indication of senescence. In some species of plants, the production of seeds and fruits causes changes in the plants causing their death. These changes are caused by chemicals that act as hormones. The chemicals cause a redirection of resources in the plants from the leaves and roots. This affects the production of seeds or fruits.

Monochasium

Monochasium is a term used in botany to refer to describe a type of inflorescence (cymose) where the branch gives rise to another branch. This is noted in plants like the buttercup and forget me not. For the cymes, they have one lateral flower. The branching could also originate from beneath the terminal hence the scorpoid inflorescence (Weitemier et al., 2014).

Monocot

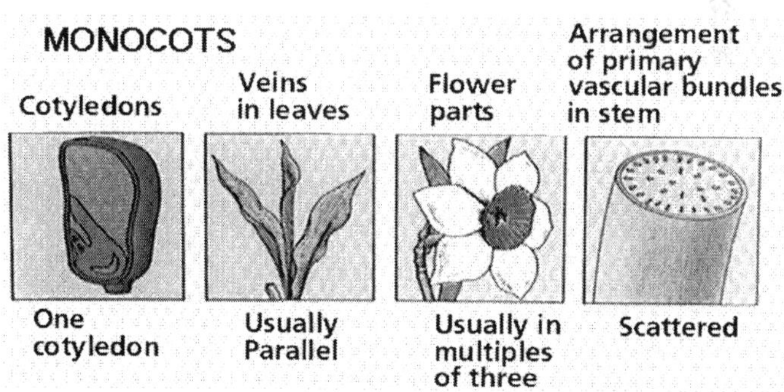

Figure 123: Monocots are plants with one cotyledon.

Source: https://www2.estrellamountain.edu/faculty/farabee/biobk/BioBook-PLANTANATII.html

Monocots are seeds with one seed leaf. The seed leaf is also called the cotyledon. Monocots are among the classification of plants. They are either dicots or monocots. Monocots can be identified by their unique characteristics. The stems of monocots have scattered vascular bundles. The stems of monocots have a large number of vascular bundles at the outside edge of the stem. The vascular bundles are surrounded by large parenchyma. Monocots do not have any pith regions. The roots of monocots have their vascular bundles arranged in a ring. The leaf veins of monocots leaves are arranged parallel to each other. They also have a parallel leaf venation. The flowers of monocots are in parts of three or multiples of three.

Monocotyledon

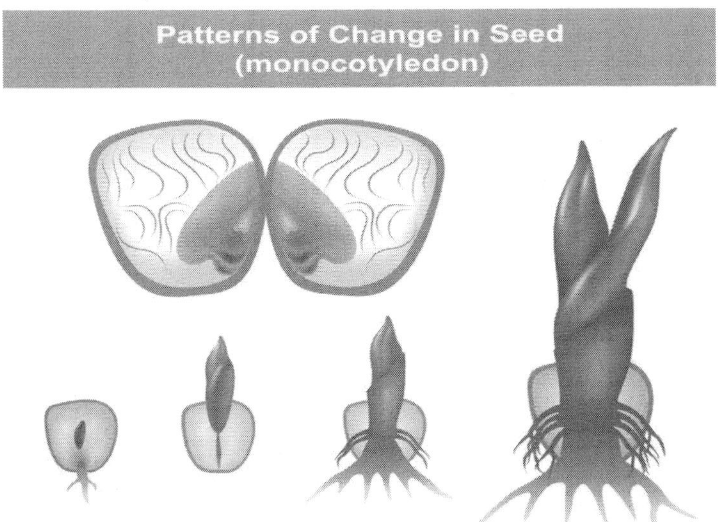

Figure 124: Monocotyledons have a single cotyledon.

Source: https://depositphotos.com/vector-images/monocotyledon.html

Monocotyledon is one of the flowering plants. They are also called monocots. They are plants that have only one cotyledon. There are about 60,000 species of monocotyledonous plants. Monocotyledons are characterized by the presence of one seed lead. The embryo is also located in the seed. The opposite of monocotyledons is the eudicotyledons that have two cotyledons. Monocotyledons have parallel-veined leaves or a parallel leaf network. They have an adventitious roots system, have no cambium and scattered vascular bundles in the stem. The flowers of monocotyledons come in threes or multiples of three. The pollen grains characteristically feature a single furrow.

Monoecious

Monoecious is a term used to refer to plants having female and male flowers on the same plant. It also refers to plants with flowers on every plant having both female and male components. A banana plant is a good example of a monoecious plant as it contains both female and male flowers (Wilkinson et al., 2012). Banana plants develop one large inflorescence having rows of female and male flowers. For a squash plant, half the flowers bloom as half of the flowers are male and the other half female. Some monoecious plants bear perfect flowers having both female and male parts in the same flower. A good example of perfect monoecious plants is the lily.

Monoicous

Monoicous is a term used to refer to haploid plants with a sexual system where both the eggs ad sperms are produced in the same gametophyte. Its opposite is diocious where the gametophyte can only produce eggs or sperms but not both together. Monoicous plants have gametophytes that produce gametes through mitosis in the gametangia. This results in the production of eggs and sperms having a similar genetic identity with their parent gametophyte. Monoicous plants could be grown in dry habitats where the ability of plants to produce sporophytes is linked to water availability.

Monograph

Monograph is the term used when a group of plants undergoes a comprehensive treatise presenting synthesis and an analysis of taxonomic knowledge of the given taxon. This results in the generation of a full account possible of the genus, tribe and family of the plants (Waraich et al., 2011). A monograph is used worldwide in evaluating and range of all taxonomic treatments of a given taxon. It includes studying the evolutionary relationships with other related ecological, palaeobotanical, morphological, genetic, cytological and taxa studies.

Monomorphic

Monomorphic is a term used in biology in describing a species in which its two sexes are phenotypically indistinguishable. This is in the absence of genetic polymorphism or sexual dimorphism. Polymorphism is the presence of two or more distinct forms in the same species. It is well illustrated by a cactus having monomorphic areoles. They have vegetative or reproductive areoles of the same shapes and are found in the same position on the tubercles

Monophyllous

Monophyllous is a term used to refer to a plant having a single leaf.

Monopodial

Monopodial is a term used to refer to a form of growth exhibited in vascular plants. Monopodial growth is characterized by the plant growing upward from a single point. In such plants, leaves are added to the apex every year. It is also noted that the stem grows longer accordingly. A common characteristic of monopodial plants is that they have a single stem or trunk. Orchids are examples of plants that exhibit monopodial growth. Orchids that exhibit monopodial growth are known to produce copious areal roots. The aerial roots tend to hang down in long drapes (Van Dijk et al., 2021). The roots have green chlorophyll underneath the grey root coverings. The green chlorophyll is used as an addition to the photosynthetic organs. These orchids do not have pseudobulbs or rhizomes. For this reason, species that grow in dry periods will have fleshy succulent leaves instead. Flowers will generate from the stem between the leaves. Other species of plants with monopodial growth may have stems forking into two. This is not necessary for monopodial orchids as the forking is not needed to ensure continued growth. This may not be the case for orchids with sympodial growth.

Monostromatic

Monostromatic is a term used to refer to plants of the genus of marine green algae that are single-layered. They are in the Monostromataceae family. An example of algae under this genus is the *Monostroma kuroshiense*. This alga is an edible product commercially cultivated in South America and East Asia. They produce two main sushi wraps namely the hirohano-hitoegusa nori and hitoegusa-nori. The *Monostroma nitidum* polysaccharides undergo agarase digestion resulting in the production of Monostroma oligosaccharides with a degree of polymerization 6. These polysaccharides have been shown to have an effective prophylactic agent against viral infection called Japanese encephalitis. This is according to *in vivo* and *in vitro* tests. Monostroma has been used in the production of sulfated oligosaccharides that seem to be promising candidates for further development as antiviral agents.

Monothecous

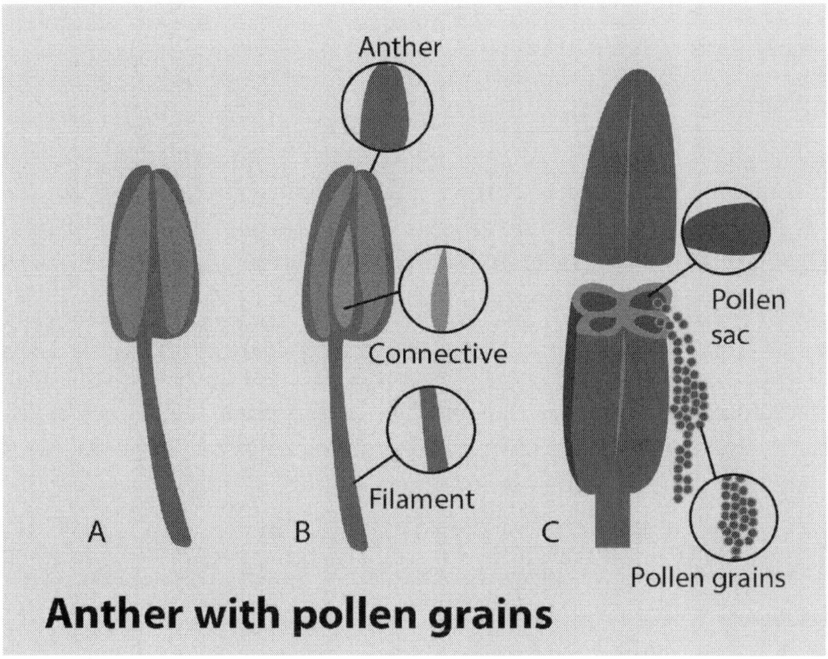

Figure 125: Monothecous anthers have single lobes.

Source: https://www.quora.com/Why-is-an-anther-called-a-tetrasporangiate

Monothecous is a term used to refer to male flowers whose anthers have a single lobe. The Malvaceae family is known to have monothecous anthers. The anthers make up the androecium which is the male reproductive part of the flower. Androecium is a group of stamen. The monothecous anther is made of single lobes with two pollen sacs. Plants with monothecous anthers also have solitary axillary inflorescence, an epicalyx, bracts, five petals and pentamerous, hypogynous or actinomorphic flowers.

Monotypic

Monotypic is a term used to refer to genera of plants made up of only a single species. The monotypic plant genera are made of four sub-categories namely the monotypic gymnosperm genera, monotypic fern genera, monotypic bryophyte genera and monotypic angiosperm genera. Other than the sub-categories there are monotypic groups. They include the division Ginkgophyta for which the only extant species is the *Ginkgo biloba*. Under the family Cephalotaceae, plants are of the genus Cephalotus and the only extant species is *Cephalotus follicularis* which is the Australian pitcher plant.

Morphology

Morphology is a term used in biology to refer to the physical and external structure of the plant. This brings about the name phytomorphology which is the study of the external structure and physical form of plants. The morphology of plants is used in visually identifying plants. Plant anatomy is very useful in describing the development of the plant (Van Bruggen et al., 2016). According to plant morphology, plants will produce new structures and tissues from the meristems. Meristems are found at the tips of organs allowing their growth. The morphology of the plant influences its growth characteristic as well as its reproductive characteristics. The morphology of the plant also describes the plants' adaptations.

Multiple fruit

Multiple fruit is fruit body formed from a cluster of fruit flowers. Multiple fruits are also called collective fruits. Fruits are produced from the inflorescence of each flower. The mass is called an infructescence after the plant flowers. Examples of multiple fruits include jackfruit, osage-orange, mulberry, pineapple and fig. After fertilization, flowers develop into drupes. The drupes expand becoming connate into multiple fleshy fruits. Multiple fruit in plane trees are formed from multiple achenes. Multiple flowers in mulberry form one fruit. The same applies to the breadfruit. In figs, the multiple fruits are in inverted inflorescence through syconium.

Mycelium

Mycelium is a mass of branching thread-like hyphae and forms the vegetative part of fungus-like bacteria colony. Fungal colonies are made up of mycelium and are found on soil. Typically a single spore will germinate to form a monokaryotic mycelium. This mycelium cannot reproduce sexually. The joining of the mycelium bodies results in the formation of fruiting bodies such as mushrooms (Van Breusegem et al., 2001).

Mycobiont

Mycobiont is a term used in biology to refer to the fungal component of a lichen. The fungal component is made up of the mycobiont cells and is part of the cells of the fungus. Mycobionts are part of the morphological forms of the lichen.

N

Natural hybrid

Figure 126: A natural hybrid is a plant that has undergone natural hybridization.

Source: https://www.anbg.gov.au/cpbr/cd-keys/orchidkey/html/intro-c_hybrid.html

Natural hybrids are plants that have undergone natural hybridization which is a significant hybridization process that occurs in some plants. Natural hybrids are progeny that are able to exploit new habitats and form new species of plants. There are few species of plants that are natural hybrids. For instance there are few natural hybrids of Australian orchids. Several natural hybrids are those of the epiphytic and terrestrial species. Studies on natural hybrids mostly focus on the natural hybrids and not of the parents producing them. There are groups of natural hybrids namely the interspecific hybrids and the intraspecific hybrids. For the interspecific hybrids, the hybrids are due to hybridization between species within a given genus. In this case, the parent species may appear to be related and may therefore share some similarities. In some cases, the parents may be distantly related or have some distinct morphology. Some of the factors that influence the production of natural hybrids include the growth habits of the plants (Van der Kooi & Ollerton, 2020). For instance, natural hybrids between native orchid species are a result of parents growing in close proximity or the flowers overlapping. Another factor is that the plant species tend to share a pollinating agent even if it is for a very short period in which there is a transfer of pollen grains. A large number of natural hybrids are noted where there is an abundance of parent species in a given location which increases chances of natural hybridization. Also, a large number of natural hybrids are also found in areas that have experienced mass flowering. Events such as good seasonal rains and summer bush fires trigger natural hybridization.

Nectar

Figure 127: Nectar is mostly found on insect-pollinated plants.

Source: https://sfyl.ifas.ufl.edu/lawn-and-garden/nectar-plants/

Nectar is produced by nectarines or nectaries. It is a sugar-rich liquid produced by flowering plants. They are found in flowers and they attract pollinating animals. Nectar is also found in extrafloral nectaries. Nectar provides animal mutualists with nutrients. The animal mutualists in turn provide herbivore protection. Some of the animals that consume nectar include bats, honeyeaters, hummingbirds, moths, butterflies, bees, wasps, hoverflies and mosquitoes. Nectar has been very useful in the evolution of nectar-eating species and the foraging economics (Twidwell et al., 2013). This is well demonstrated by the evolution of the African honey bee where nectar foraging behavior has greatly influenced its divergent evolution. Nectar is an important part substance in the production of honey as it is the sugar source of honey. Other than the production of honey, nectar is also useful in horticulture and agriculture as some adult stages of some predatory insects feed on nectar. The parasitoid wasps depend on nectar as their primary source of food. The wasps in turn hunt agricultural pest insects as food for their young. Flowers that frequently secrete nectar are often visited by pollinators. After pollination has occurred, the remaining nectar is reabsorbed into the plant.

Nectary

Nectary is a part of the plant involved in the production of nectar. A nectary consists of an annular disk that surrounds the ovary. A nectary is a specialized nectar-producing structure of a flower. Nectaries (plural for nectary) produce nectar which is a solution of one or more sugars and other compounds. Substances produced by nectaries often at times function as rewards for animals in a bid to promote animal pollination. A nectary may be pad-like in that it develops as a discrete pad of tissue that extends only part-way around the base of the flower. Floral disks that consist of doughnut-shaped or disk-like mass of tissue that surrounds the ovary base or top may function as a nectary. The nectariferous disks may be intrastaminal, extrastaminal or staminal to the androecium. This makes it easy for the androecium to deposit pollen on the body of the pollinator as it feeds on nectar. The perigonal nectary is found on the perianth and is found at the base of tepals, petals or sepals. On the other hand, the septal nectaries are specialized tissues embedded within the septa of an ovary and secretes nectar via a pore at the apex or the ovary base.

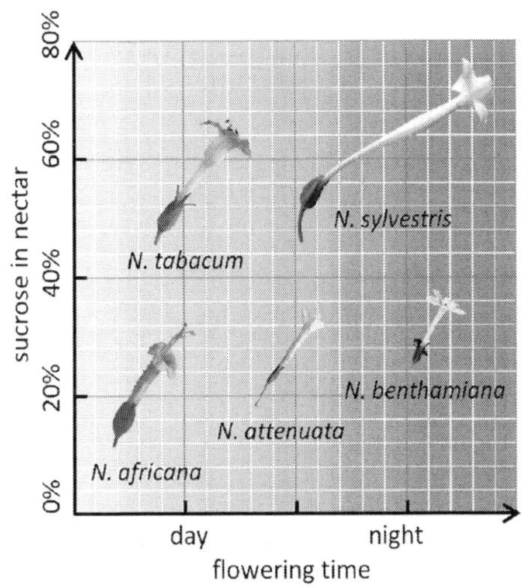

Figure 128: A nectary is involved in the production of nectar.

Source: https://www.frontiersin.org/research-topics/6575/new-perspectives-on-the-biology-of-nectaries-and-nectars

Neophyte

Neophyte is a term used in botany to refer to a plant species that is not native to a geographical region. This group of plant species was recently introduced. There are non-native plants that are long established in an area called archeophytes.

Nerve

Nerve is a term used to refer to the nerve plant also called *Fittonia albiveins*. The nerve plant is a flowering plant in the family Acanthaceae. It is native to the rainforests of northern Brazil, Ecuador, Bolivia, Peru and Colombia. The Nerve plant is an evergreen perennial and is characterized by its dark green foliage with strong contrasting red or white veins. The Nerve plant is at times called the mosaic plant. The nerve plant is grown as a houseplant in temperate regions where the temperature falls below 50 °F. The Nerve plant is a creeping evergreen perennial that can grow to a height of 15 cm (Taiz et al., 2015). They also have lush green ovate leaves and accented veins of white to deep pink. Their stems have a short fuzz. In some nerve plants, some small buds may be noted where the stem splits into leaves. Other types of nerve plants have a carmine-red nervature. Nerve plants have small flowers with white or off-white color. When grown as house plants, the nerve plants should be watered regularly failure to which the plant will "faint".

Figure 129: The nerve plant is also called the mosaic plant.

Source: https://littleprinceplants.com/our-plants/plant-solutions/beau-chateau/fittonia-zebrano-nerve-plant/

Node

Nodes are areas on stems where buds are found. Nodes are usually sites characterized by great cellular growth and activity. It is the area where small buds develop into flowers, stems and leaves. Plant notes are usually identified when pruning the plant. This is done to avoid damaging of the nodes. Pruning is normally done above the nodes while leaving a distance between the pruning site and the node. Pruning is done to encourage the buds at the node to begin the development process which eventually leads to the formation of new leaves and stems. Internodes are areas between two nodes. The length of an internode is influenced by many factors such as genetics.

Reduced soil fertility also affects internode length by casing a reduction in the length. This is solved by applying high-nitrogen fertilizer to soil. The absence of light causes an increase in the internode length though it causes a spindly stem. The spindly stem is a situation known as etiolation or stretch. Seasons also influence the internode length. Plants that grow in the early-season tend to have long internodes while those that grow in the late season have short internodes (Szpak, 2014). The plants' energy can also influence the length of the internode. The internode tends to be short if the plants' energy is diverted into development and growth. Internode length is also affected by exposure to herbicides and plant growth regulators.

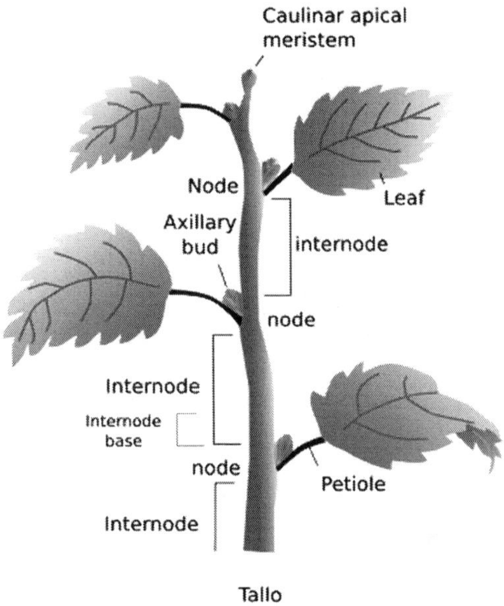

Figure 130: The distance between nodes is the internode.

Source: https://mmegias.webs.uvigo.es/02-english/2-organos-v/guiada_o_v_tallo.php

Nomenclature

Nomenclature is the formal scientific method of naming living organisms. The scientific naming of plants is called the botanical nomenclature. Although nomenclature is distinct from taxonomy it is still related to it. While plant taxonomy is concerned with classifying and grouping plants, botanical nomenclature gives the name of the plant after plant taxonomy is done. The Linnaeus' Species Plantarum is the starting point for modern nomenclature. The International Code of Nomenclature for algae fungi and plants governs the botanical nomenclature. The body is a replacement for the International Code of Nomenclature for algae (Stenberg, 2017). The code of nomenclature also covers fossil plants. Botanical nomenclature also differs from plant taxonomy in that it is the body of rules that prescribes the name that applies to the given taxon. It also determines whether a new name may be coined. Nomenclature has been useful in that it enables easy identification of plant species. It also helps classify newly found plants. It is useful in identifying whether a plant is harmful or harmless to both human beings and animals. It also helps to categorize fossil plants.

Noxious

Noxious is a term used in botany to refer to the noxious weed. The noxious weed is a weed that has been identified by agricultural or governing authorities as a plant that can injure a horticultural or agricultural crop, ecosystems, natural habitat, livestock or human beings. Noxious weeds may also be called injurious weeds or harmful weeds. Noxious weeds may find their way into an ecosystem by accident, mismanagement or ignorance though some noxious weeds are native to some areas. Noxious weeds exhibit aggressive growth, multiply quickly without natural control and may cause adverse effects when ingested or when it comes into contact with the skin. In several parts of the world, noxious weeds are a great problem. They affect areas of nature reserves, forest management and agriculture. Noxious weeds have found their way to new countries and regions through contaminated shipments of crop seeds and feeds. In some areas, they were intentionally introduced as ornamental plants. A good example of noxious plants is the ragwort. Ragworts are known to produce copious amounts of nectar. They have been very useful in ensuring the survival of other pollinators. They also act as habitats and larval host foods. The wild parsnip *Pastinaca sativa* is a noxious weed in the USA. It is used to provide large tubular stems creating space for bee species to hibernate in. They are also larval food for the swallowtail butterflies.

Figure 131: *Pastinaca sativa* and ragworts are examples of noxious weeds.

Source: https://www.thespruce.com/pictures-of-noxious-weeds-2132944

Nucellus

The nucellus is the central part of a mature ovule. The nucellus carries the embryo sac having eight nuclei. The nuclei have a set of chromosomes hence the name haploid nuclei. Two nuclei are located near the center and are called the polar nuclei.

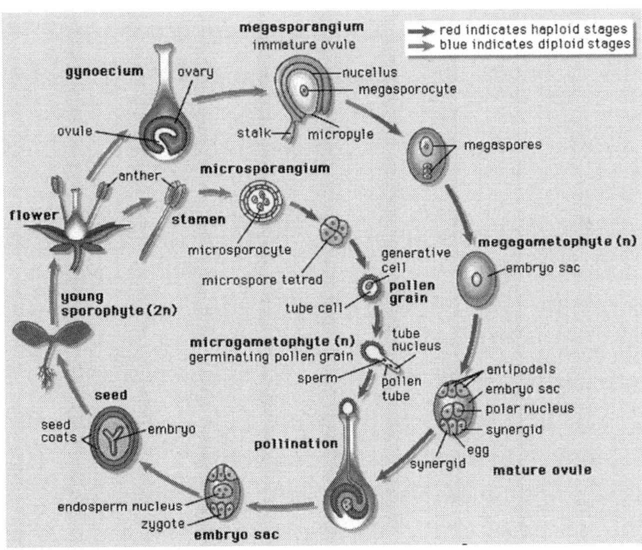

Figure 132: The nucellus carries the embryo sac.

Source: https://www.britannica.com/science/nucellus

The nucellus develops as the ovule matures. The nucellus plays a vital role in the development of the ovule into a seed, a process that is dependent on the fertilization of a plant that comes after the plant is pollinated. Usually, the pollen grain will land on the upper surface of the pistil and germinate. This results in the production of pollen tubes that grow down within the style towards the ovule. The nucellus carries the embryo sac and the nucleus of the ovary. The pollen tube grows and penetrates the wall of the nucellus and finds its way to the nucleus allowing fertilization of the ovules (Singh, 2015).

Nut

Figure 133: Walnuts, chestnuts, hazelnuts and acorns are examples of nuts.

Source: https://www.scientificamerican.com/article/aflatoxins-in-nuts-danger-or-hype/

Nuts are fruits consisting of tough or hard nutshells that protect the kernel which is usually edible. Nuts are generally used more so in the culinary sense. The term nuts is at times used to refer to a variety of dry seeds. In the biological context, the term nut is used to refer to seeds whose shell does not open to release the seed. Naturally, it is expected that seeds from fruits are able to free themselves from the shell. This does not apply to nuts such as acorns, chestnuts and hazelnuts (Schuurink et al., 2006). These nuts have a hard shell wall. They also originate from the compound ovary. The term, however, is not restricted to seeds whose shells do not break. Brazil nuts, walnuts, pistachios, pecans and almonds are nuts but not in the botanical sense. The term nut is used to refer to any hard-walled edible kernel. Nuts are known to have high levels of nutrients and energy. It is a good source of food.

Nutlet

A nutlet is a small nut. Like other nuts, nutlets may have hard nutshells protecting their kernel and may be edible. Nutlets are also used in the culinary world. Nutlets mas originate from the compound ovary.

O

Obconic

Obconic is a term used in botany to refer to a cone with an inverted shape. The term is mostly used when referring to fruits with hypanthium structure whose apical end is attached to the stem usage of the term is less used when referring to the pistil structure. The term is also used when referring to fungi when referring to the ascospore. Currently, the term obconic is used in botany to refer to an entire plant. It can be used to refer to the shape of an entire shrub. Among the plants known to have an obconic structure is the hypanthium which is a western USA plant. One of its sub species has obconic structures. For the hypanthium plant, the obconic characteristic is a subspecies diagnostic and determinant. The Toyon shrub is a species of hypanthium with the obconic shape (Smith & Knapp, 2001).

Obcordate

Obcordate is a term used to refer to a leaf that is heart-shaped with a notch apical. The leaf can also be described to have a reversed cordate shape. Though the leaf is heart-shaped, it is attached to the stalk by the pointed end

Obdiplostemonous

Obdiplostemonous is a term used to refer to plants with two sets of stamens with the inner series alternating with them while the outer series is opposite to the petals. The term Obdiplostemonous is formally used when describing species, genera and families. The Obdiplostemonous plants produce flowers that have twice as many stamens as petals. In the flower, the stamens are arranged in two rings with those in the inner having gaps between the outer ones while those in the outer rings are opposite the petals. There are about 20 plant families known to bear such kinds of flowers. They include geraniums, the sorrels, oxalis and heathers. The term can be broken down to better explain how it was derived. Ob-represents opposite, diplo-represents doubled or twofold, stemon-represents thread and ous-is the ending. The term therefore can be described as relating to two opposite placed sets of stamens.

Oblanceolate

Oblanceolate is a term used to refer to a leaf with a rounded apex and tampering base. In most cases, the leaf is a simple leaf that has not been divided into parts. For plants having such leaves, the upper region of the stem has a characteristic square-shaped cross-section with long internodes which separate the whorls of elliptical to the oblanceolate leaves. An example of a plant that bears oblanceolate leaves is *Crassula orbicularis*. When growing plants for shade gardens *Crassula orbicularis* is known for its wonderful appearance. It is easy to grow. It is a succulent perennial plant that can grow up to a height of 0.25 m high when they flower. The plant has a shallow root system. They also grow in rosettes of few to several leaves in some cases, they produce runners. Other than the leaves being oblanceolate, they are fleshy, flat, narrow, wide and elliptic. The plant produces a terminal, rarely spike-like inflorescence and leaf-like buds. The flowers of these plants are cup-shaped, produce a must scent and they are rarely blackish when in the bud. The flowers have petals that are generally whitish and are about 2-5 mm long. The flowers are at times brownish-red.

Obligate

Obligate is a term used to refer to a photoperiodic plant that requires a long or short night before flowering starts. The long or short night could also stimulate germination in the plant. Some of these plants grow in wetlands and are called obligate wetlands. These plants grow under natural conditions in wetlands. Alpine violet, Switchgrass, Poison ivy, Red maple, Ground ivy, Christmas fern, Virginia creeper, White clover and White pine are examples of obligate upland plants. Water horsetail, royal fern, soft-stem bulrush, wooly sedge, cattails, pickerel weed, water lily and duckweed are examples of obligate wetland plants.

Oblique

Oblique is a term used to refer to the leaves of plants. The bases of such leaves are unequal, slanting with one side being wider, larger or rounder than the other. Examples of plants with oblique leaf bases are the Common Hackberry and the American Elm.

Obloid

Obloid is a term used to refer to plants with a three-dimensional oblong shape.

Obovate

Obovate is a term used in botany to refer to leaves that are egg-shaped and have a narrow end at the base. Obovate leaves are simple leaves that have not been divided into parts.

Orbicular

Orbicular is a term used in botany to refer to leaves that are circular or almost circular in shape. Quaking Aspen, the Canberry Cotoneaster and Arrowwood Viburnum are examples of plants with orbicular leaves (Rusman et al., 2019).

Order

Order is a term used in biological classification in a taxonomic rank. It is used when classifying and recognizing organisms by nomenclature codes. The taxonomic rank follows a descending order which is: life, domain, kingdom, phylum, class, order, family, genus and species. In most cases, the order fits between class and family. The superorder is the immediately higher rank. In some cases, the superorder is at times placed directly above the order. The suborder is added directly beneath the order. The order is also a taxonomic unit in the rank of order. The plural of order is orders. The taxonomist determines the type of plant that does or does not belong to a given order. They also determine whether a given order should be recognized. In some cases, taxonomists may not exactly agree as there are no hard rules to be followed when describing and recognizing a given order.

Ortet

Ortet is a term used in botany individual plants that give rise to clones of plants by propagation or vegetative reproductive.

Orthotropous

Orthotropous is a term used in botany to refer to upright and straight plants with the micropyle at the apex.

Oval

Oval is a geometrical term that is used to refer to any curve resembling an ellipse an egg. The shape of an egg can be oval or ovoid.

Ovary

Ovary is a term used in botany that refers to an enlarged basal portion of the pistil. The pistil is the female organ of a flower. The ovary carries ovules. The ovules develop into seed after fertilization. When the ovary after fertilization will become the fruit. They fruit can either be fleshy or dry. The fruit enclosed the seeds. There are different types of ovaries including the unicarpellate ovary also called the simple ovary (Reid et al., 2009). The simple ovary is formed from a simple carpel which is an evolutionarily modified leaf. The simple ovary has one locule and it carries a number or ovules. The other kind is the multicarpellate ovary that has several carpels. They may also have one or more locules. When classifying a plant, the position of the ovary is used. A superior ovary is one in which the ovary is attached above the floral part while an inferior ovary is one in which the ovary is attached below the floral parts.

Ovate

Ovate is a term used in botany when referring to the shape of a leaf. The widest point of an ovate leaf is below the middle. The length to wide ratio usually determines the kind of ovate leaves. The first category of ovate leaves has a length to width ratio of 1.5: 1 which is less than 2:1. The second category is that of widely ovate = length to width ratio from 1.2:1 that can be less than 1.5:1 and the final category is that of a narrowly ovate = length to width ratio of 2:1 going as low as 3:1.

Ovoid

Ovoid is a term used to refer to leaves. There are several species of trees known to have ovoid leaves more so shrubs. The American snowbell is an example of a tree with ovoid leaves. Other than being ovoid the leaves are also deciduous and about 3-6 cm long with protruding veins. The margins of the leaf are also serrated the botanical name of the plant is *Styrax americanus*. Another tree that has ovoid leaves in the Grey Alder also called *Alnus incana*. The ovoid leaves of the Grey alder are tapered while the leaf edges are roughly double serrated. The underside of the leaf is gray felty pilose. The leaves are about 10 cm long.

Ovule

Ovule is a term used to refer to a plant structure that develops into a seed after fertilization. When an ovule matures it contains food tissues that are covered by one or two seed coats. The food tissues are called integuments. Ovules have a small opening called the micropyle. The integuments allow the pollen tube to enter (Poff & Zimmerman, 2010). This allows the discharge of sperms nuclei. They find their way into the embryo sac which is a large oval cell. It is where both fertilization and development occur. The base of the ovule is attached to the stalk which bears it. For gymnosperms such as allies and conifers, the ovules lie uncovered in the scale of the cone while in flowering plants one or more ovules could be enclosed in an ovary. The ovary develops into the fruit.

P

Pachycaul

Pachycauls are plants having a trunk that is excessively thick for their height and few branches. The name is derived from the Greek pachy-, which means thick or stout, and the Latin caulis, which means stem.

Figure 134: Fouquieria columnaris.

Source: https://en.wikipedia.org/wiki/Pachycaul

Pachycladous

(Botany) Having disproportionately thick stems.

Palea

Palea is the name given to one of the spikelet's bract-like organs of the Poaceae family.

In botany, a spikelet is the usual arrangement of the flowers of grasses, sedges, and a few other Monocots.

The palea is the bigger of two chaff-like bracts that surround the grass floret (the other being the lemma). It is frequently split at the tip, indicating that it is a bifurcated structure generated by the union of two separate organs. This has led to suspicion that it is what is left of the grass sepals (outer perianth whorl), namely the two adaxial members of the three-membered whorl typical of monocots (Pauchard et al., 2009).

According to differing botanical interpretations, the third member may be absent or represented by the lemma.

The expression of MADS-box genes in palea during development, as seen in eudicot plant sepals, lends support to the perianth explanation.

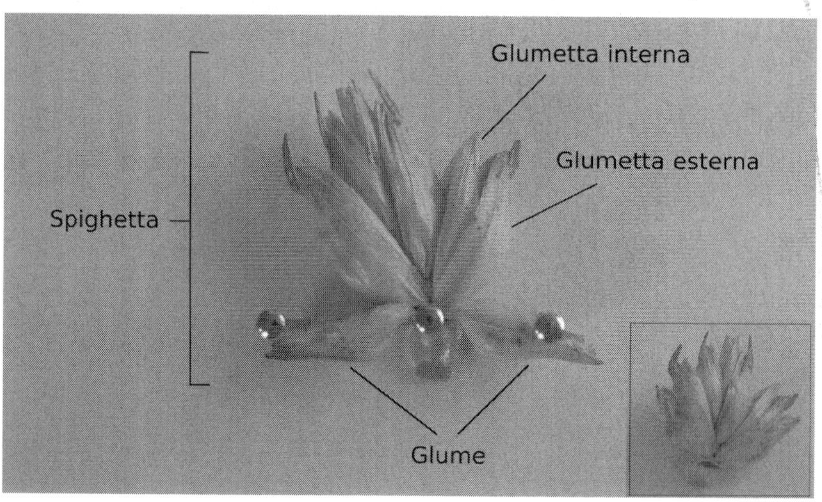

Figure 135: Parts of a single grass spikelet, consisting of two glumes, four fertile florets, with one additional central floret that may or may not be sterile.

Source: https://en.wikipedia.org/wiki/Spikelet

Palmate

Palmate (botany adjective) [PAHL-meyt]: A compound leaf with three or more lobes or leaflets that all sprout from the same point at the end of a stem. Palmate leaves typically have their leaflets arranged in the shape of fingers on a hand. Virginia Creeper is one of the High Park plants with palmate leaves! This vine has a woody stem and palmate leaves with 5 leaflets.

This plant may go overlooked by passers-by in the spring and summer since it fits in with the many other green plants flourishing in the park. However, when the season of change arrives, this vine makes a strong appearance! Virginia creeper is one of the most beautiful fall color displays, becoming a rich warm color when the temps drop. The hue might vary from brilliant red to reddish-orange to burgundy.

Figure 136: Palmate leaves.

Source: https://highparknaturecentre.com/index.php/blog/2019/word-of-the-week-palmate

Virginia creeper is frequently confused with Poison Ivy, another woody vine that thrives in the park. Both Poison Ivy and Virginia creeper have palmately compound leaves, but one way to tell them apart is the number of leaflets—Poison Ivy has three, while Virginia Creeper has five. If the

Virginia creeper has lost two of its leaflets, the identification can become more difficult.

Palmatifid

Of a leaf shape; palmately cleft rather than lobed. Compound - composed of more than one part; "compound leaves are composed of several lobes; "compound flower heads"

Palmatisect

Of a leaf cut into lobes to more than halfway in a palmate form (Prajapati & Modi, 2012).

Pandurate

(Of plant leaves) shaped like the body of a fiddle.

Papilionate

- A butterfly-like flower with a pea-like corolla.
- Flowers with the characteristic irregular and butterfly-like corolla seen in many, but not all, plants of the species-rich Faboideae subfamily of legumes are known as papilionaceous flowers (from Latin: papilion, a butterfly). Flores papilionacei, according to Tournefort, was named after Valerius Cordus, who applied it to bean blooms.

Pappus

In Asteraceae flower heads, the pappus is a modified calyx, or part of an individual floret, that covers the base of the corolla tube. Although such structures are not formed by the flower's calyx, the term is widely applied for similar structures existing as a wind-dispersal technique for seeds of other plant families, such as some Apocynaceae genera.

The bristles (sometimes feathery), awns, scales, or absence of bristles, awns, scales, or absence of bristles, awns, scales, or absence of bristles, awns, scales, or absence of bristles, awn The pappus is too tiny to view without magnification in some species. The feathery bristles of the pappus operate as a "parachute" in species such as Dandelion and Eupatorium, allowing the seed to be borne by the wind.

The name is derived from the Ancient Greek word pappos, Latin pappus, which means "old man," and is applied to a plant with bristles (assumed to be an Erigeron species) as well as the woolly, hairy seed of certain plants (Neumann et al., 2009).

The pappus of the dandelion is essential for wind-aided seed dissemination. The pappus's flight is stabilized, and greater lift and drag are produced, by producing a divided vortex ring in its wake. The pappus also has the ability to modify its shape in response to moisture in a variety of ways that aid in germination. The change in form might influence the pace of abscission, allowing for increased or decreased germination depending on the favorable conditions.

Figure 137: The pappus-clad fruits that make up the familiar "dandelion clock" being dispersed by the wind (family Asteraceae).

Source: https://en.wikipedia.org/wiki/Pappus_(botany)

Paraperigonium

The non-reproductive component of the flower is the perianth (perigonium, perigon, or perigone in monocots), and a perigone is a structure that produces an enclosure around the sexual parts, comprised of the calyx (sepals) and the corolla (petals) or tepals. Perianth is made up of the Greek words peri (around) and anthos (flower), whereas perigonium is made up of peri and gonos (seed, sex organs). In mosses and liverworts (Marchantiophyta), the perianth is the sterile simple tubular tissue that covers the female reproductive organ (or developing sporophyte).

In some plants (e.g,, Narcissus, Passiflora (passion flower), some Hippeastrum, and Liliaceae), the corona (paraperigonium, paraperigon, or paracorolla) is a ring or pair of adaxial cellular extensions emerging from the corolla or the exterior boundary of the stamens. It's most common where the corolla lobes protrude from the corolla tube.

Figure 138: Flower of Narcissus showing an outer white corolla with a central yellow corona (paraperigonium).

Source: https://en.wikipedia.org/wiki/Perianth

Parenchyma

Plant ground tissue encompasses all tissues that are neither dermal nor vascular. It is classified into three categories based on the composition of the cell walls.

- Parenchyma Cells
- Collenchyma Cells
- Sclerenchyma Cells

Parenchyma is a multipurpose ground tissue that serves as "filler" tissue in soft plant components. It is responsible for the cortex (outer region) and pith (center part) of stems, the cortex of roots, the mesophyll of leaves, the pulp of fruits, and the endosperm of seeds, among other things. Parenchyma cells are frequently alive and may remain meristematic at maturity, which means they can divide if stimulated (Nikunj et al., 2014). They feature thin and flexible cellulose cell walls, are polyhedral when close-packed, but can be roughly spherical when isolated from their neighbors. The cells of the parenchyma are normally big. They have huge central vacuoles that allow the cells to store and control ions, waste products, and water. Tissue specialized for food storage is commonly formed of parenchyma cells.

Parietal

Connected to the peripheral walls of a structure, such as ovules connected to placentas on the ovary's wall.

Parietales is a flowering plant order. It is a precise botanical word for the plants' placentae parietales (parietal placentation). As a consequence, it might still be utilized today (for a taxon above the rank of family). The ending -ales is merely a coincidence equivalent to that of the rank of order.

Paripinnate

The configuration of feather-like or multi-divided characteristics growing from opposite sides of a common axis is known as a pinnation (also known as a pennation). Pinnation can be found in biological morphologies, crystals such as ice or metal crystals, and patterns of erosion or stream beds.

Paripinnate leaves are pinnately complex leaves with leaflets produced in pairs down the rachis rather than a single terminal leaflet; sometimes known as "even-pinnate."

Imparipinnate: pinnately complex leaves with a single terminal leaflet rather than a pair of terminal leaflets; also known as odd-pinnate.

Figure 139: Paripinnate or even pinnate.

Source: https://en.wikipedia.org/wiki/Pinnation

Pauciflor

A plant that has only a few flowers per inflorescence.

Pectinate

Pinnately split, having narrow segments placed tightly together like comb teeth.

Pedate

In biology, a pedate formation is one that mimics or has the quality of feet. It is derived from the Latin verb "pedo," which means "to equip with feet." (Novello et al., 2020).

In botany, the term applies to compound leaves, veins, or other structures whose parts radiate from a central point (as in a palmate structure), but the lateral divisions are additionally split in two. More generally, it can be used to designate a complex leaf with a terminal leaflet and branched axes on either side that curve out and back, and to which leaflets are attached on the curve's outer side.

Figure 140: A pedately divided leaf.

Source: https://en.wikipedia.org/wiki/Pedate

Pedicel

A pedicel is a stem that connects the inflorescence to a single flower. Pedicellate inflorescences are the name given to such inflorescences.

Pedicel refers to a structure linking a solitary flower to its inflorescence. Flowers are classified as sessile when they lack a pedicel. Pedicel is also referred to the stem of the infructescence. The term "pedicel" is derived from the Latin pediculus, which means "little foot." A peduncle is a stem or branch from the main stem of the inflorescence that holds a bunch of pedicels. A pedicel may be accompanied by a bract or bracts.

Peduncle

In botany, a peduncle is a stalk that holds an inflorescence or a single flower, or, after fecundation, an infructescence or a single fruit. Bracts may be present on the peduncle at nodes (a form of cataphyll). Above the peduncle, the rachis is the main axis of an inflorescence. The peduncle lacks flowers, while the rachis is festooned with them (Mercuri et al., 2002).

A scape is a peduncle that grows from the earth, either from a compacted aerial stem or a subterranean stem (rhizome, tuber, bulb, corm, etc.), with few or no bracts save around the rachis or receptacle.

The pedunculate oak's acorns are carried on a lengthy peduncle, hence the tree's name.

Peltate

Shield-like, with the stalk attached to the lower surface and not to the margin.

Pendulous

Hanging, for example, an ovule attached to a placenta on the top of the ovary. Compare suspended.

Pentamerous

A pedicel is a stem that connects the inflorescence to a single flower. Pedicellate inflorescences are the name given to such inflorescences (Matyjaszczyk & Śmiechowska, 2019).

Pedicel refers to a structure linking a solitary flower to its inflorescence. Flowers are classified as sessile when they lack a pedicel. Pedicel is also referred to the stem of the infructescence. The term "pedicel" is derived from the Latin pediculus, which means "little foot." A peduncle is a stem or branch from the main stem of the inflorescence that holds a bunch of pedicels. A pedicel may be accompanied by a bract or bracts.

In nature, five or three parts per whorl have the highest frequency of occurrence, but four or two parts per whorl are not uncommon.

Pepo

- A type of berry formed from an inferior ovary and containing many seeds, usually large with a tough outer skin (e.g. a pumpkin or cucumber).
- An indehiscent fleshy one-celled many-seeded berry (such as a pumpkin, squash, melon, or cucumber) that has a hard rind and is the characteristic fruit of the gourd family

Perennating

From season to season, in relation to organisms that survive vegetatively. It is common for there to be a period of diminished activity between seasons.

Perennial

Perennial plants, sometimes known as long-lived plants, live for more than two years. The expression is typically used to identify a plant from annuals and biennials, which have shorter lives. The term is also used to distinguish vegetation with little or no woody development (secondary growth in girth) from trees and shrubs, both of which are technically perennials (Mandel, 2019).

Herbaceous perennials are blooming plants that grow and flower in the spring and summer, disappear in the autumn and winter, then return in the spring from their rootstock or other overwintering components. A plant that is a perennial in its native habitat or in a warmer garden may be handled as an annual by a gardener and planted out every year, from seed, cuttings, or divisions, based on the harshness of the local environment (temperature, moisture, organic content in the soil, microbes).

Tomato vines, for instance, can thrive for several years in their native tropical/subtropical environment but are grown as annuals in temperate areas since their above-ground bulk does not survive the cold.

There are also evergreen, or non-herbaceous, perennials that maintain a mantle of leaves all year, such as Bergenia. Subshrubs, such as Penstemon, are a form of transitional plant that preserves a vestigial woody structure in the winter.

Perfoliate

With its base wrapped around the stem (so the stem appears to pass through it), as in leaves and bracts.

Perianth

The non-reproductive component of the flower is the perianth (perigonium, perigon, or perigone in monocots), and a perigone is a component that produces an enclosure around the sexual organs, consisting of the calyx (sepals) and the corolla (petals) or tepals. In mosses and liverworts (Marchantiophyta), the perianth is the sterile tubelike tissue that surrounds the female reproductive organ (or developing sporophyte) (Lindström et al., 2013).

The perianth of flowering plants can be dichlamydeous/heterochlamydeous, which means that the calyx and corolla are separate, or homochlamydeous, which means that they are alike (and the sepals and petals are collectively referred to as tepals). The perianth is said to be biseriate when it has two whorls.

While the calyx might be green, as in sepaloid, it can also be vividly colored, as in petaloid. When undifferentiated tepals resemble petals, they are referred to as "petaloid," as in petaloid monocots, which are orders of monocots with brightly colored tepals. They are also known as lilioid monocots because they include Liliales. Pollinators are attracted to the corolla and petals, although this may be augmented by more specialized elements like the corona.

Pericarp

Simple fruits are produced by a single ovary and might contain one or many seeds. They can be both meaty and dry. During the development of fleshy fruit, the pericarp (ovary wall) and other accessory components become the fleshy section of the fruit. Berries, pomes, and drupes are examples of fleshy fruits.

In other fruits, such as mangosteen or pomegranate, the edible section is derived from the aril rather than the ovary, and in pineapple, tissues of the flower and stem produce food. Grass grains are single-seed simple fruits with the pericarp and seed coat united into a single layer. Caryopsis is the scientific name for this variety of fruit. Cereal grains such as wheat, barley, and rice are examples (Luan et al., 2020).

The pericarp of fleshy fruits is normally composed of three separate layers: the epicarp (also known as exocarp), which is the outermost layer; the mesocarp, which is the middle layer; and the endocarp, which is the inner layer around the ovary or seeds. The peel of a citrus fruit is made up of the epicarp and mesocarp. The layers of the pericarp are not easily visible in dry fruits.

Pericycle

The pericycle is a cylinder of parenchyma or sclerenchyma cells that resides just within the endodermis and is the most outer component of a plant's stele.

Although it is made up of non-vascular parenchyma cells, it is nevertheless regarded as a part of the vascular cylinder since it, like the vascular tissues it surrounds, emerges from the procambium.

It is also capable of producing lateral roots in eudicots. This basic meristem tissue gives rise to branch roots. The pericycle contributes to the vascular cambium commonly branching into a cork cambium in plants during secondary development.

Certain chemicals from the endodermis and surrounding vasculature are delivered to the pericycle in angiosperms, where they encourage the formation of root meristems.

The pericycle is found in plant roots between the endodermis and the phloem. It is found in dicot stems around the ring of vascular bundles in the stele.

The pericycle in dicot roots strengthens the roots and protects the vascular bundles.

The vascular cambium in dicot roots is entirely secondary in origin, originating from a piece of pericycle tissue (Lambers et al., 2008).

By quickly dividing near the root's xylem components, the pericycle governs the production of lateral roots.

It has been observed that it is frequently confused with other components of the plant. Its distinct ring structure, on the other hand, makes it more easily recognized.

Attempts to isolate such tissue in the past have proved effective. Monocot roots rarely branch, but when they do, it is from the pericycle.

Perigonium

Synonym of perianth in flowering plants.

In mosses, the leaves that encircle the antheridia, also known as a splash-cup, for example, in *Polytrichum juniperinum*.

Perigynium

A sac formed by a stretched tubular bract, or a utricle when closed completely, that surrounds the pistillate flower of sedges.

Perigynous

An ovary is a female reproductive organ of the flower, also known as a gynoecium, in flowering plants. It is the portion of the pistil that contains the ovule(s) and is positioned above, below, or at the point of attachment with the petals and sepals. Because the pistil can be made up of a single carpel or multiple fused carpels (e.g., dicarpel or tricarpel), the ovary can also be made up of a single carpel or many joined carpels. The style and stigma are placed over the ovary and are where pollen falls and germinates before passing through the style and fertilizing one particular ovule for each pollen grain. The ovaries of some wind-pollinated plants have greatly shrunk and changed (Klassen, 2001).

A half-inferior ovary (also known as "half-superior," "subinferior," or "partially inferior") is embedded or surrounded by the receptacle. This occurs in the flowers of plants in the Lythraceae family, which contains crape myrtles. Perigynous or half-epigynous blossoms are what these flowers are called. Some schools do not identify half-inferior ovaries and instead lump them in with either the superior or inferior ovaries.

A half-inferior ovary contains approximately equal amounts of ovary above and below the insertion location. Other fractions might be used to illustrate different levels of inferiority. For example, a "one-fourth inferior ovary" has around one-fourth of its length under the insertion point. Likewise, above the insertion, just one-quarter of a "three-quarters inferior ovary" is visible.

Perule

- The scales that cover a leaf or flower bud, or a scale-like leaf that surrounds the bud. "Naked" buds are those that lack perulae.
- In Camellias, the terminal bracts and sepals merge and are referred to as perules.
- In certain orchids, a type of sac created by the attached bases of the two lateral sepals.

Petal

Petals are modified leaves that enclose a flower's reproductive components. To attract pollinators, they are frequently brilliantly colored or strangely shaped. The corolla is the collective name for a flower's petals. Petals are generally accompanied by a second set of modified leaves known as sepals, which comprise the calyx and lay slightly under the corolla. The perianth, or non-reproductive section of a flower, is made up of the calyx and the corolla (Katova, 2007). When the petals and sepals of a flower are difficult to discern, the petals and sepals are referred to as tepals. Genera such as Aloe and Tulipa are examples of plants where the term tepal is applicable. Rosa and Phaseolus, on the other hand, have distinct sepals and petals. Petaloid refers to undifferentiated tepals that resemble petals, as in petaloid monocots, orders of monocots with brilliantly colored tepals.

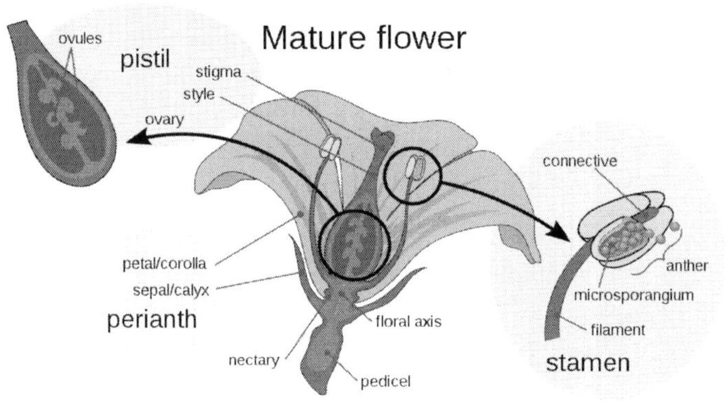

Figure 141: Parts of a mature flower.

Source: https://en.wikipedia.org/wiki/Petal#/media/File:Mature_flower_diagram.svg

Because they include Liliales, they are also known as lilioid monocots.

Although petals are typically the most visible element of animal-pollinated flowers, wind-pollinated plants, such as grasses, have extremely few petals or none at all (apetalous).

Petaloid

Like a petal; soft in texture and colored conspicuously.

Petiole

The petiole is the stalk that connects the leaf blade to the stem and allows the leaf to be rotated to face the sun. This affords the plant a unique leaf configuration. In some species, stipules are outgrowths that grow on both sides of the petiole. The term petiolate alludes to leaves with a petiole, whereas sessile or apetiolate refers to leaves without a petiole. In entomology, the phrase is referred to as a petiole (insect anatomy)

A petiole is a stem that connects a leaf to a plant's stem. In petiolate leaves, the leaf stalk can be long, as in celery and rhubarb leaves, short, or totally lacking, in which case the blade links directly to the stem and is said to be sessile. Subpetiolate leaves are sessile and have a very tiny petiole. The broomrape family Orobanchaceae is an instance of a species with always sessile leaves. Petiolate and sessile leaves may exist in separate species in several other plant groupings, such as the speedwell genus Veronica.

Figure 142: Leaf of dog rose (*Rosa canina*), showing the petiole, two leafy stipules, the rachis, and five leaflets.

Source: https://en.wikipedia.org/wiki/Petiole_(botany)

Petiolule

The leaflets of plants with compound leaves are linked to an extension of the petiole known as the rachis. A slender stalk called the petiolule connects each leaflet to the rachis. Pulvina are swelling patches at either end of the petiole that are made of a flexible tissue that helps the leaf to move. Pulvina are found in the bean family Fabaceae as well as the prayer plant family Marantaceae. A pulvinulus is a pulvinus on a petiolule (Jeger et al., 2007).

Phaneranthous

Showy, as in showy flowers that advertise to pollinators, as opposed to aphananthous (unshowy)

Phanerogam

Plants that generate seeds, also known as phanerogams or phaenogams, are classified as spermatophytes, hence the alternative name seed plants. They are a subgroup of land plants known as embryophytes.

The term phanerogams or phanerogamae is derived from the Greek phanerós, which means "transparent," as opposed to the cryptogamae kruptós, which means "hidden," and the suffix gamé, which means "to marry." This terminology discriminated between plants that had hidden sexual organs (cryptogamae) and those that had apparent sexual organs (phanerogamae).

Phenology

Phenology is the study of periodic events in biological life phases and how they are affected by seasonal and interannual variations in climate as well as environmental circumstances (such as elevation).

Good examples include the dates of leaf and flower creation, the first flight of insects, the first appearance of migratory birds, the date and time of deciduous tree leaf coloration and fall, and the timing of temperate-zone honey bee colonies' developmental cycles. In the scientific research on ecology, the phrase is used more widely to designate the time range for any seasonal biological occurrence, such as the dates of the last occurrence (e.g., the seasonal phenology of a species may be from April through September).

Because many of these events are particularly sensitive to variations in climate, particularly temperature, phenological data can be a useful proxy for temperature in historical climatology, especially in the study of anthropogenic global warming (Hirschberg, 2001). In Europe, for example, viticultural data of grape harvests have been used to reconstruct a record of summer growing season temperatures stretching back over 500 years. In addition to having a broader historical background than quantitative measures, phenological data provide high temporal resolution of ongoing changes associated with global warming.

Phloem

The living tissue of vascular plants known as phloem transfers the soluble organic material generated in photosynthesis referred to as photosynthates, notably sucrose, to various parts of the plant. Translocation is the term used to describe this kind of movement. The name phloem is derived from the Ancient Greek word phloiós, meaning "bark." It refers to the lowest layer of bark in trees. In 1858, Carl Nägeli invented the expression.

Phloem tissue is made up of conducting cells, also known as sieve elements, parenchyma cells (which include both specialist companion cells or albuminous cells and unspecialized cells and supporter cells like fibers and sclereids), and sclereids.

Figure 143: Phloem (orange) transports products of photosynthesis to various parts of the plant.

Source: https://en.wikipedia.org/wiki/Phloem

Photobiont

A photobiont is a photosynthetic partner of lichens. Lichen photobionts are made up of a wide range of bacterial and eukaryotic species. The bulk of lichens have a photobiont that is either a green alga (Chlorophyta) or a cyanobacterium. Some lichens exhibit both types. Algal photobionts are known as phycobionts, whilst cyanobacterial photobionts are known as cyanobionts. Around 100 photosynthetic partners from 40 genera and 5 different families have been discovered to interact with the lichen-forming fungi (prokaryotic: Cyanophyceae; eukaryotic: Trebouxiophyceae, Phaeophyceae, Chlorophyceae).

Common algal photobiont genera include Trebouxia, Trentepohlia, Pseudotrebouxia, and Myrmecia. Trebouxia is the most common genus of green algae found in lichens, contributing to around 40% of all lichens. The term "trebouxioid" alludes to a photobiont that is presumably in the genus Trebouxia or closely mirrors a member of that genus and hence is likely a member of the class Trebouxiophyceae (Hadwiger, 2013). Trentepohlia is the 2nd most frequent green algae genus. Over 100 kinds of eukaryotes have been identified as photobionts in lichens. All of the algae are presumably capable of coexisting in nature and in lichen.

A "cyanolichen" is a lichen whose primary photosynthetic component is a cyanobacterium (photobiont). The majority of cyanolichens are also ascolichens, however, some basidiolichens, such as Dictyonema and Acantholichen, have cyanobacteria as a partner.

Photosynthesis

Photosynthesis is the process through which plants and other organisms convert light energy into chemical energy, which is then discharged to fuel the organism's operations through cellular respiration. Some of this biochemical energy is preserved in carbohydrate molecules such as sugars and starches, which are made from carbon dioxide and water - thus the name photosynthesis, which is taken from the Greek phs, "light," and sunthesis, "coming together." In most cases, oxygen is also generated as a byproduct, holding three times the amount of chemical energy as carbohydrates (Guo & Liu, 2007). The bulk of plants, algae, and cyanobacteria perform photosynthesis; these organisms are classified as photoautotrophs. Photosynthesis is largely accountable for producing and maintaining the level of oxygen in the Earth's atmosphere and also supplying the bulk of the energy needed for life on Earth.

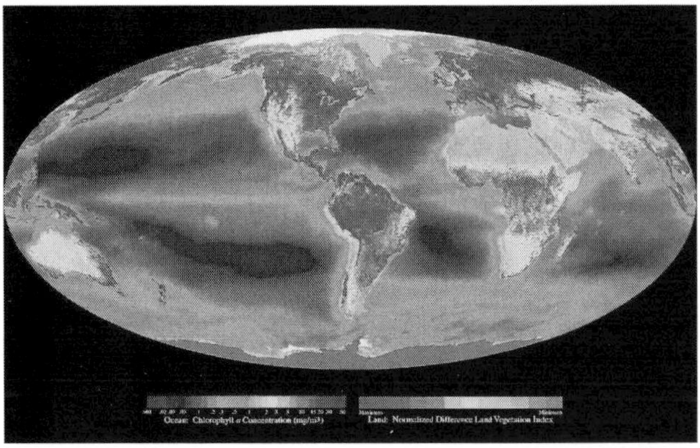

Figure 144: Composite image showing the global distribution of photosynthesis, including both oceanic phytoplankton and terrestrial vegetation. Dark red and blue-green indicate regions of high photosynthetic activity in the ocean and on land, respectively.

Source: https://en.wikipedia.org/wiki/Photosynthesis

Phyllary

A phyllary, also known as an involucral bract or tegule in botanical language, is a single bract of a compound flower's involucre. The involucre is a collection of bracts. Phyllaries are leaf-like appendages that create one or more whorls directly beneath a flower head.

Phyllid

A leaf-like extension of the stem in Bryophytes

Phyllode

Phyllodes are modified petioles or leaf stems that look and function like leaves. In some plants, they flatten and broaden, while the leaf shrinks or disappears entirely. As a result, the phyllode appears to serve the purpose of the leaf. *Euphorbia royleana*, which is cylindrical, and Opuntia, which is flattened, are two significant examples (Arruda et al., 2015).

They are prevalent in the genus Acacia, particularly in the Australian species, which were previously classified as Acacia subg. Phyllodineae. Partially developed phyllodes with decreased leaves can be seen on occasion, especially on younger plants.

Daviesia, a genus in the Fabaceae family, is distinguished by the presence of phyllodes in the plants.

Phyllopodium

(In ferns) A short offshoot of the stem that bears the frond and stays linked to the rhizome after the frond is dropped.

Phylloplane

The surface of a leaf thought to be a habitat for organisms

Phyllosphere

When seen as a habitation for microorganisms, the phyllosphere refers to the entire above-ground surface of a plant in microbiology. The phyllosphere includes the caulosphere (stems), phylloplane (leaves), anthosphere (flowers), and carposphere (fruits). Below-ground microbial ecosystems are the rhizosphere and laimosphere (i.e., the thin volume of soil surrounding root or subterranean stem surfaces). Most plants sustain a diverse variety of microorganisms, including bacteria, fungi, archaea, and protists. Some are beneficial to the plant, whereas others act as plant pathogens, inflicting harm or death to the host plant.

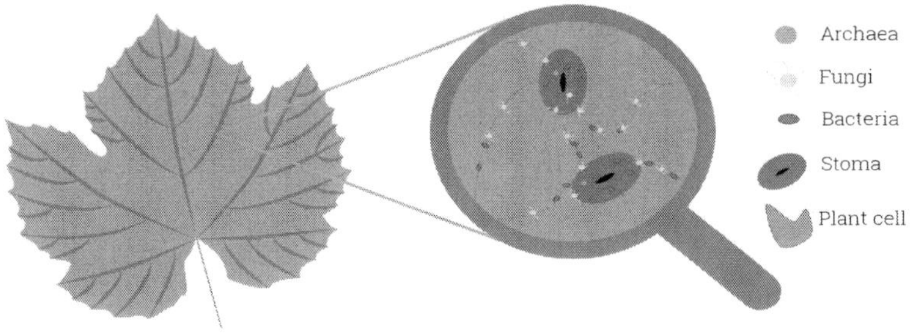

Figure 145: The plant aerial surface, mostly occupied by leaves, is inhabited by diverse microorganisms, forming the phyllosphere.

Source: https://en.wikipedia.org/wiki/Phyllosphere

Phylum

A phylum is a level of classification or taxonomic rank in biology that is below kingdom and above class. Although the International Code of Nomenclature for algae, fungi, and plants considers the terminology as equal, division has traditionally been used instead of phylum in botany (Franklin et al., 2013). The animal kingdom Animalia has roughly 31 phyla, the plant kingdom Plantae contains approximately 14, and the fungus kingdom Fungi contains approximately 8 phyla, depending on definition. Current phylogenetic study is revealing the links between phyla that are encompassed in larger clades, such as Ecdysozoa and Embryophyta.

Phytomelan

Phytomelanin (phytomelan) is a black, inert organic substance found in Asparagales and Asteraceae but uncommon in other taxonomic groups that forms a crust-like covering of some seeds. Phytomelanin may be found in the majority of Asparagales families (although not in Orchidaceae). It is mechanically hard and creates a resistant substance, however, it is flexible in the developing fruit before hardening. It looks to be a polyvinyl aromatic alcohol chemically, and it is assumed to be secreted from the hypodermis. It appears to provide resistance to insect predators as well as resistance to desiccation.

Pileus

The pileus is the scientific name for the cap, or cap-like element, of a basidiocarp or ascocarp (fungal fruiting body) that holds a spore-bearing surface known as the hymenium. The hymenium (hymenophore) on the pileus's underside might be made up of lamellae, tubes, or teeth. A pileus is a form of fungus that can be found in agarics, boletes, some polypores, tooth fungi, and ascomycetes.

Pilei can take on a variety of morphologies, which can change during the span of a fungus's growth cycle. The most common pileus shapes are hemispherical or convex pilei. Convex pilei frequently expand until they become flat. A convex pileus is seen in several well-known species, including the button mushroom, Amanita species, and boletes.

Some mushrooms, such as the parasol mushroom, have distinguishable bosses or umbos and are known as umbonate. An umbo is a chunky protrusion in the cap's center. Some fungi, such as chanterelles, resemble funnels or trumpets. The pileus is referred to as infundibuliform in these circumstances (Foxcroft et al., 2017).

Pinna

The term "frond" refers to a big, split leaf. In both common English and botanical nomenclature, the leaves of ferns are referred to as fronds, and some botanists restrict the term to this class. Other botanists use the term frond to describe the broad leaves of cycads, palms (Arecaceae), and a number of other flowering plants such as mimosa and sumac.

A pinna (plural pinnae) is a leafy part of the blade of a pinnate (feather-shaped) frond, a petiolule is the stalk bearing the pinna, and a costa is the primary vein or mid-rib of the pinna (plural costae).

A frond that has been divided into pinnae is referred to as once pinnate. In some fronds, the pinna is further divided into segments, resulting in a bipinnate frond. Pinnules are the segments into which each pinna is divided, whereas rachillae are the appendages of the rachis that hold these pinnules. In rare situations, a frond may be tripinnate, in which case the pinnule divides are alluded to as final segments.

Pinnate

The organization of feather-like or multi-divided characteristics growing from opposite sides of one axis is known as a pinnation (also known as a pennation). Pinnation can be found in biological morphologies, crystals such as ice or metal crystals, and patterns of erosion or stream beds.

The phrase comes from the Latin word pinna, which means "feather," "wing," or "fin." "Pectination," which is a comb-like arrangement of pieces, is a comparable idea. Pinnation is used to distinguish it from "palmation," where the components or structures split out from a singular point. The words "pinnation" and "pennation" are cognates, and while they are sometimes used differently, there is no continuous difference in their meaning or usage.

Pinnation is a botanical term that describes the creation of multiple independent structures (like leaflets, veins, lobes, stems, or extensions) along a common axis. Pinnately compound leaves, for example, are leaf blades that have been divided once and have leaflets on both sides of a rachis. Pinnately divided leaves are seen on many palms, particularly feather palms, and the bulk of cycads and grevilleas. The fronds of most ferns are pinnate or more finely divided, and the leaflets or sections of ferns are typically called "pinnae" (singular "pinna"). Plants with pinnate leaves are usually known as "feather-leaved." The Glossary of Botanical Terms is used to define the majority of the terms in this section (Bourgaud et al., 2001).

Pistil

The gynoecium can have single or multiple different pistils. A pistil usually comprises an elongated basal component known as the ovary, an elongated portion known as the style, and an apical structure known as the stigma that takes pollen.

The ovary (from Latin ovum, literally meaning "egg") is the bigger basal part of the egg that contains placentas and tissue ridges that hold one or perhaps more ovules (integumented megasporangia). Ovules and/or placentas can be formed on gynoecial structures or, less regularly, on the flower apex.

A locule is the compartment where ovules develop (or sometimes cell).

The style is a pillar-like stalk with pollen tubes leading to the ovary. Some blooms, such as Tulipa, do not have a separate style, and the stigma is linked directly to the ovary.

The style of some plants, such as lilies, has a hollow tube or has transmitting tissue whereby the pollen tubes grow (Chandler & Sanchez, 2012).

The stigma is usually found towards the tip of the style, which is the area of the carpel(s) that collects pollen (male gametophytes). It is frequently sticky or feathery to capture pollen.

The word "pistil" comes from the Latin word pistillum, meaning "pestle." A pistillode is a male flower's sterile pistil.

Pistillate Flower

A flower containing one or more pistils but no stamens. Sometimes called a female flower.

Pistillode

A sterile or rudimentary pistil.

Pith

Pith, also known as medulla, is a tissue found in the stems of vascular plants. Pith is made up of soft, spongy parenchyma cells that can retain starch in some situations. Pith is found in the core of the stem of eudicotyledons. It also spreads into flowering stems and roots in monocotyledons. The pith is surrounded by a ring of xylem, which is surrounded by a ring of phloem.

While fresh pith growth is often white or pale in color, the tissue commonly darkens to a deeper brown tone as it grows. Pith is normally present in initial growth in trees, although it is frequently replaced - in large part - by xylem in the trunk and older branches. The pith at the middle of the stem of some plants may dry out and disintegrate, resulting in a hollow stem.

Some plants, like as walnuts, have unique chambered pith with multiple short cavities. Cells on the pith's periphery may develop differently than cells in the rest of the pith in some plants. This layer of cells is referred to as the pithamus' perimedullary area. An example of this can be found in the ivy Hedera helix (Craigie, 2011).

The term pith is also applied to the pale, porous inner layer of the peel of citrus fruits (such as lemons) and other hesperidia, which is more correctly referred to as mesocarp or albedo. The term comes from the Old English word pia, which signifies substance, and is related to the Middle Dutch pitte (modern Dutch pit), which means a fruit pit.

Plant Breeders Rights (PBR)

Plant breeders' rights (PBR), also known as plant variety rights (PVR), are rights granted to the breeder of a new variety of plant that grant the breeder exclusive authority over proliferating material (including seed, cuttings, divisions, tissue culture) and harvested content (cut flowers, fruit, foliage) of the new variety for a set period of time.

With these privileges, the producer has the choice of becoming the variety's exclusive marketer or licensing it to someone else. To apply for these exclusive rights, a variety must be original, distinct, uniform, and stable. A variety is

- new if it has not been commercialized in the country of protection for more than one year;
- distinct if it differs from all other known varieties by one or more important botanical characteristics, such as height, maturity, color, and so on;
- uniform if the plant characteristics are consistent from plant to plant within the variety.
- Plant features are said to be stable if they are genetically fixed and hence remain the same from generation to generation, or after a cycle of reproduction in the case of hybrid types.

The breeder must also assign an approved "denomination" to the variety, which serves as its general name and must be used by anybody who markets the variety (Clemens et al., 2013).

Plant variety rights are often given by national offices after an assessment. The plant variety office receives the seed and grows it for one or more seasons to ensure that it is distinct, stable, and uniform. If these tests are passed, exclusive rights are awarded for a set amount of time (usually 20/25 years for trees and vines, or 25/30 years for trees and vines). To keep the rights, annual renewal fees are required.

Plumule

The plumule is the portion of a seed embryo that develops into the stalk that bears the plant's first genuine leaves. The plumule in most seeds, such as the sunflower, is a tiny conical structure with no leaf structure. The plumule does not develop until the cotyledons have grown above ground. This is known as epigeal germination. A leaf structure is apparent on the plumule in the seed of some seeds, such as the broad bean. The plumule grows up through the earth, but the cotyledons remain below the surface, resulting in the development of these seeds. This is referred to as hypogeal germination.

Pluriflor

Having many flowers per inflorescence.

Pluriovulate

Having many ovules as in placentae, carpels, or ovaries.

Pneumatophore

Aerial roots are roots above the ground. They are commonly adventitious. Pneumatophores are unique aerial roots that allow plants to breathe air in wet soil settings. The roots can develop downward from the stem or upward from normal roots. If they emerge from the earth, some botanists define them as aerating roots rather than aerial roots. The epidermis of these roots is coated with lenticel (small pores) that suck in air and turn it into spongy tissue, which subsequently distributes oxygen throughout the plant via osmotic channels as needed. Pneumatophores separate the Black and Grey mangroves from other mangrove species (Fahad et al., 2017).

Southeast Asian fishermen mold the pneumatophores of *Sonneratia caseolaris* (aka “Mangrove Apple”) into small floats to make corks for fishing nets.

Taxodioideae members produce woody terrestrial appendages known as cypress knees that sprout from their roots. Initially, these structures were supposed to function as pneumatophores, but current research has found no evidence of this.

Figure 146: The Grey Mangrove (*Avicennia marina*)’s pneumatophorous aerial roots.

Source: https://en.wikipedia.org/wiki/Aerial_root#/media/File:Pneumatophore_overkill_-_grey_mangrove.JPG

Pod

- A legume, the fruit of a leguminous plant, a dry fruit of a single carpel, splitting along two sutures.
- A siliqua and silicula, the fruit of Brassicaceae, a dry fruit composed of two carpels separated by a partition.

Podocarpium

The genus Podocarpium is in the family Leguminosae in the major group Angiosperms (Flowering plants).

Pollen

Pollen is a powdery substance made up of pollen grains, which are seed plant microsporophytes that create male gametes (sperm cells). Pollen grains have a strong sporopollenin coat that protects gametophytes as they migrate from the stamens to the pistil of flowering plants or from the male cone to the female cone of coniferous plants. When pollen settles on a suitable pistil or female cone, it germinates and produces a pollen tube that transports the sperm to the ovule carrying the female gametophyte (Elmer & White, 2018). Individual pollen grains are so minute that they require magnification to be seen in detail. Palynology is the study of pollen, which is immensely important in paleoecology, paleontology, archaeology, and forensics. Pollen is used in cross-pollination to transport haploid male genetic elements from the anther of one flower to the stigma of the other.

Pollen-Mass

Pollen-grains cohering by a waxy texture or fine threads into a single body; pollinium, e.g. in orchids.

Pollen Transmitting Tissue

The tissue in the style of a flower through which the pollen tubes grow.

Pollination

Pollination is the movement of pollen from an anther (male part) of a plant to the stigma (female part) of a plant, allowing fertilization and seed formation. Pollination is most commonly accomplished by an animal or by wind. Animals such as insects, birds, and bats, as well as water, wind, and even plants themselves, when self-pollination happens within a closed bloom, are pollinating agents. Within a species, pollination is common. In nature and in plant breeding, pollination between species can result in hybrid progeny (De Waele & Elsen, 2007).

In angiosperms, the pollen grain (gametophyte) germinates on the stigma and creates a pollen tube that extends down the style till it hits an ovary. Its two gametes travel down the chute to the carpel, where the gametophyte(s) containing the female gametes reside.

One male nucleus unites with the polar organs to form the endosperm tissues after crossing through the micropyle, while the other unites with the ovule to create the embryo. As a result, the term "double fertilization" came into use. This method would result in the development of a seed that would contain both nutritive tissues and an embryo.

Figure 147: Female Xylocopa with pollen collected from night-blooming cereus.

Source: https://en.wikipedia.org/wiki/Pollination

The ovule in gymnosperms is not contained in a carpel, but is exposed on the surface of a specialized support organ, such as the scale of a cone, making carpel tissue penetration unnecessary. The process differs based on the gymnosperm division in question. Gymnosperms have two basic fertilization methods. Cycads and Ginkgo have motile sperm that swims directly to the egg inside the ovule, but conifers and gnetophytes have sperm that cannot swim and must be delivered to the egg via a pollen tube.

Polygamodioecious

Some plants have bisexual and male flowers, while others have bisexual and female blooms. Androdioecious, andromonoecious, dioecious, monoecious, polygamomonoecious, and polygamous are all synonyms for androdioecious.

Polygamomonoecious

The presence of male, female, and bisexual flowers on the same plant.

Polystemonous

Having a large number of stamens; the number of stamens should be at least twice the number of sepals or petals, but not necessarily three or four times that amount (Chen & Yamaguchi, 2002).

Pome

A pome is a type of fruit produced by flowering plants of the subtribe Malinae of the Rosaceae family. Pomes that are well-known include the apple, pear, and quince.

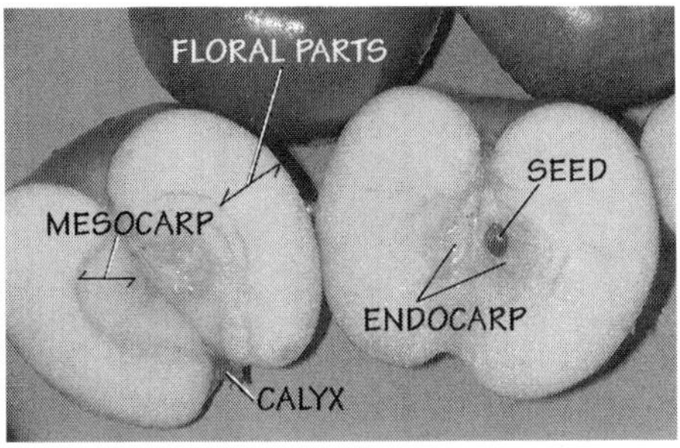

Figure 148: An apple is a pome fruit. The parts of the fruit are labeled.

Source: https://en.wikipedia.org/wiki/Pome

Prickle

A hard, pointed protrusion from a plant's surface (involving several layers of cells but lacking a vein); a sharp outgrowth from the bark, removable without ripping wood (Chen & Yamaguchi, 2002).

Primary Vein

The single vein or cluster of veins in a leaf that is noticeably larger than the others. The single primary vein in pinnate venation is usually situated in the middle of the leaf; in palmate venation, numerous similar veins radiate from a point at or near the leaf's base.

Prophyll

A prophyll is a leaf-like structure, similar to a bracteole, which extends beneath a solitary flower or pedicel. The lowest bract on a peduncle can also be referred to as a bract.

Procumbent

Prostrate spreads along the ground but does not root at the nodes; not as near to the ground as prostrate.

Propagule

In biology, a propagule is any material that facilitates an organism's propagation to the next phase of its life cycle, like dispersion. In appearance, the propagule is often distinct from the parent organism. Plants (through seeds or spores), fungi (via spores), and bacteria all produce propagules (for example endospores or microbial cysts).

In disease biology, pathogens are said to produce infectious propagules, which are the components that spread a disease. Microorganisms present in host material include bacteria, viruses, fungi, and protists. Infectious influenza propagules, for example, are conveyed in particles of host saliva or mucus expelled during coughing or sneezing.

A propagule is any plant material used in horticulture for plant propagation. In asexual reproduction, a propagule is typically a stem cutting (Zehnder et al., 2007).

In some plants, a fragment of a leaf or a portion of a root might be used. In sexual reproduction, a propagule is a seed or spore. Any part of the plant can be used in micropropagation, which is a type of asexual reproduction, though it is typically a highly meristematic component like root and stem ends or buds.

Prostrate

Rooting at nodes that touch the soil surface when lying flat on the ground.

Protandrous

- Having male sex organs that mature before female sex organs, for example, a flower discharging pollen before the stigma is receptive.
- Individual flowers in plants are dichogamous if their function separates the two sexes in time, even if the plant as a whole may have functionally male and functionally female blooms open at the same time. A flower is protogynous if it has a female function first, then a male function, and protandrous if it has a male function first, then a female function. This was supposed to prevent inbreeding, although it could be a more general method for minimizing pollen-pistil interference.

Proteranthous

Including fresh foliage coming before flowers.

Prothallus

A prothallus, also known as a prothallium, is the gametophyte stage of the life of a fern or other pteridophyte. The phrase is also sometimes used to designate the immature gametophyte of a liverwort or peat moss (Yamada et al., 2003).

The prothallus develops from a germinating spore. It is a short-lived and modest heart-shaped formation that is typically 2–5 millimeters wide, with a number of rhizoids (root-like hairs) sprouting beneath and the sex organs archegonium (female) and antheridium (male) developing beneath (male). The appearance of various species varies substantially. Some are green and photosynthesis occurs, but others are colorless and feed subterranean as saprotrophs.

Protogynous

- Having female sex organs that mature before male sex organs, as in a flower discharging pollen after the stigma has stopped being receptive.
- In plants, individual flowers are called dichogamous if their function has the two sexes separated in time, although the plant as a whole may have functionally male and functionally female flowers open at any one moment. A flower is protogynous if its function is first female, then male, and protandrous if its function is male then female. It used to be thought that this reduced inbreeding, but it may be a more general mechanism for reducing pollen-pistil interference.

Pseudanthium

A pseudanthium (Greek meaning "false flower") is a flower-like inflorescence. The term is sometimes applied to structures that are neither real flowers nor true inflorescences. Flower heads, composite flowers, and capitula are all types of pseudanthia, which are inflorescences [where anything from a small bunch to hundreds or even thousands of blooms are grouped collectively to form a unified flower-like construct]. Pseudanthia can manifest itself in a variety of ways. The actual flowers (the florets) are usually small and severely reduced, but the pseudanthium itself can sometimes be rather massive (as in the heads of some varieties types of sunflower).

Figure 149: What appear to be individual flower "petals" are actually individual entire ray flowers, with a dense pack of individual tiny disc blooms in the center. The collection of flowers in the head of this sunflower is called a pseudanthium or a composite because it has the overall look of a single flower.

Source: https://en.wikipedia.org/wiki/Pseudanthium

Pulvinus

A pulvinus (plural pulvini) is a joint-like swelling at the base of a plant leaf or leaflet that allows for non-growth-dependent movement (nyctinastic and thigmonastic). Pulvini are members of various plant families, including the Fabaceae (Leguminosae) bean family and the Marantaceae prayer plant family.

Pulvini are situated around the petiole's base or apex, as well as where the leaflets of a complex leaf are placed in the rachis. They are composed of a vascular tissue core surrounded by a malleable, dense cylinder of parenchyma cells with thin walls. A geniculum is another name for a pulvinus (Yakhin et al., 2017).

Pulvinar movement is caused by changes in turgor pressure, which causes the parenchyma tissue to contract or expand. The process starts when sucrose is released from the phloem into the apoplast. The increasing sugar concentration in the apoplast lowers the water potential, causing potassium ions to be effluxed from the surrounding cells.

This is followed by a water outflow, culminating in an abrupt change in turgor pressure in the pulvinus cells. Aquaporins on the pulvini vacuole membrane allow for water efflux, which contributes to the change in turgor pressure. The principle is comparable to that of stomatal retraction.

Pyrene

A pyrena or pyrene (also known as a "pit" or "stone") is a fruitstone formed within a drupe or drupelet by the ossification of the fruit's endocarp or lining. It is made up of a firm endocarp tissue that surrounds one or more seeds (also called the "kernel"). The hardened endocarp that makes up the pyrene forms a physical barrier around the seed, protecting it from diseases and herbivory.

While many drupes are monopyrenous, meaning they contain only one pyrene, pome-type fruits with a hard, stony (rather than leathery) endocarp are polypyrenous, meaning they contain numerous pyrenes.

Pyriform

Pyriform refers to solid things that are essentially conical in shape, having the broadest end at one end and the narrowest end at the other. The distal third of their length is typically the broadest, whereas the proximal end, the base, where the stalk, if present, joins, is typically the smallest.

Pyrophiles

Pyrophile refers to plants that require fire for reproduction.

The passage of fire is required to raise the dormancy of seeds of pyrophile plants such as Cistus and Byblis, an Australian passive carnivorous plant, by rising temperature and producing smoke.

Imperata cylindrica is a Papua New Guinea plant. Even though it is green, it easily ignites and causes fires on the slopes.

Pyrophytes

Pyrophytes are plants that have developed to withstand fire.

Some species benefit from fire. "Passive pyrophytes" withstand the impacts of fire, especially when it moves swiftly, and can thus outcompete less resistant plants that are injured. "Active pyrophytes" have a comparable competitive advantage to passive pyrophytes, but they additionally produce volatile oils, which increase the occurrence of beneficial fires. Pyrophile plants are those that require fire to complete their reproductive cycle.

Q

Queen Anne's lace

Also known as Daucus Carota, the Queen Anne's lace plant grows up to 5 feet tall. This plant has finely cut leaves that resemble ferns and clusters of white flowers that bloom along a lengthy stem. It is a delicate perennial flower native to Europe and some parts of North America and Asia. Queen Anne's Lace has a reputation for attracting butterflies, bees and hummingbirds - which is one of the reasons it was selected U.S. State Flower of Pennsylvania in 1893. Recognize it by its many-branched wands of white flowers surrounded by green bracts that resemble lace draping from the stems. Native to North America; most often found in moist meadows, wetlands, pastures and roadsides (Yakhin et al., 2017). Queen Anne's Lace is a biennial plant and a member of the classic carrot family. Its long, slender stalks can grow to a height of 4-5' with light green, fern-like foliage. This perennial, hardy plant produces 1/8"-1/4" diameter flower clusters that appear along the stalk in late spring. It grows well in fertile, loose soil and has excellent drought tolerance once established. Queen Anne's lace is often overlooked by today's gardeners, but it gets its name from Britain's Queen Anne (1665 to 1714) who grew it for its culinary qualities.

Figure 150: Queen Anne's lace.

Source: Picture by Wikipedia

Quinine

Quinine comes from the bark of the cinchona tree that grows in the mountains of Peru and Java. It is among the most widely used drugs in modern medicine as it is the key ingredient in producing commonly-prescribed antibiotics, including chloroquine and its derivatives, along with being a key ingredient in some other common medicines. In these medicines, Quinine is used to treat malaria, to prevent migraine headaches and to treat leg cramps as well as help reduce fluid accumulation caused by congestive heart failure and help rid the body of excess fluid after treatment for cancer (Wisler & Norris, 2005). Aminoquinolines are used to create many other drugs as well. There are over 700 species of quinine plants growing in the rainforests of the Andes and these plants provide a natural source for quinine, which is an antimalarial drug. The bark of these trees is harvested and crushed to release the bitter-tasting alkaloid, quinine. Quinine is useful in treating malaria because it suppresses the reproduction of malaria parasites in red blood cells. A number of other species of the genera Cinchonae also contain quinine, but *Cinchona ledgeriana* (from which quinine is extracted) and *Cinchona calisaya* (that gives rise to alkaloid chemically similar to quinine) are considered the best sources.

Figure 151: Quinine plant.

Source: Picture by Britannica

R

Raceme

Raceme combines technology and design for a unique product. It is a modern planter and lighting system that functions as a bench, table or floor lamp when illuminated. Deep and tapered like traditional flower arrangements, the planter is kept to a shallow depth so water can drain freely. The planter is designed in such a way that excess water is kept out of the base, making it easier to clean, ideal for indoor and outdoor use. It can also be described as a type of plant which makes a flower cluster at the tip of each stalk. It is so-called because it is racemosely flowered, i.e., each plant bears many small flowers at the tips of the shoots. The cluster can be described as panicle-like and raceme-based. It is a name used to describe a hardy annual vine legume with an oblong, red-brown pod a few inches in length (Weitemier et al., 2014). Raceme (from Latin racēma, meaning a tuft of hair or plume of feathers) is a kind of floral arrangement in which the flowers are borne along an unbranched, indeterminate axis, usually lasting only one season. The raceme has a short peduncle with small flowers densely crowded at the tip. It bears no true leaves and has no stalk, nor is there any modification of the stem serving as a petiole or leafstalk. The term can also refer to a group of flowers borne on such an axis – for example, the inflorescence of some members of the squash family bear an edible raceme.

Figure 152: Raceme wild flower.

Source: Picture by Dreamstime.com

Rachilla (rhachilla)

The only known species in its genus. The Rachilla plant is a perennial, spreading from rhizomes to form large patches that can reach up to 1 m tall. When cut in half vertically, it is possible to see the very thin white paper-like tissue (the endodermis) that gives the shoots their shape (Sensitive Plants and Flowers). Three Gram-negative rod-shaped bacteria were isolated from infected root ends and identified as *Rhizobium leguminosarum* biovar trifolii. it is a climbing plant from the family Asclepiadaceae, which includes many species known by popular names such as miracle plants, pipevine species and snake fruits. Rhachilla, also commonly referred to as Snake creeper, can be grown indoors in pots, or outdoors on tree branches. It produces clusters of small white, green or red flowers above attractive grey-green leaves. It is a native vine that grows in southern Europe and Asia Minor. This plant can be seen in rocky areas, hedges and woods and mainly on limestone rocks. A rhachilla typically arises by developing from the axil of a leaf is a type of stem called a rhachis. Plants with such flower clusters are called rhachiads. They are usually herbaceous, usually perennial, and usually contain more than one flower. The related term cyme refers to a cluster or group of flowers.

Rachis

Rachis is an herb that grows in the mountains. Its small, white flowers grow in dense clusters at the tops of its rose-colored stalks. The rachis is the name that is also given for the slender axis, or central supporting column, of a compound leaf. The rachis functions to transmit stress (such as wind or weight) from the petiole to the lamina (leaf blade), and also provides attachment for the lamina to the petiole. The rachis is sometimes also referred to as an axis. Under this definition, the term rachis covers shoots, and sometimes roots (Wilkinson et al., 2012). It is distinct from the spandex of dicotyledonous plants, a band of tissue that supports the entire inflorescence. Olea rachis lilies, also known as desert lilies, or beccabunga (which means 'beaked pimple' in Italian), are a genus of bulbous perennials native to the Canary Islands as well as part of the Macaronesian islands of Azores and Madeira. They grow best in Mediterranean climate zones and can survive extreme heat. The flowers are used by entomologists to attract butterflies.

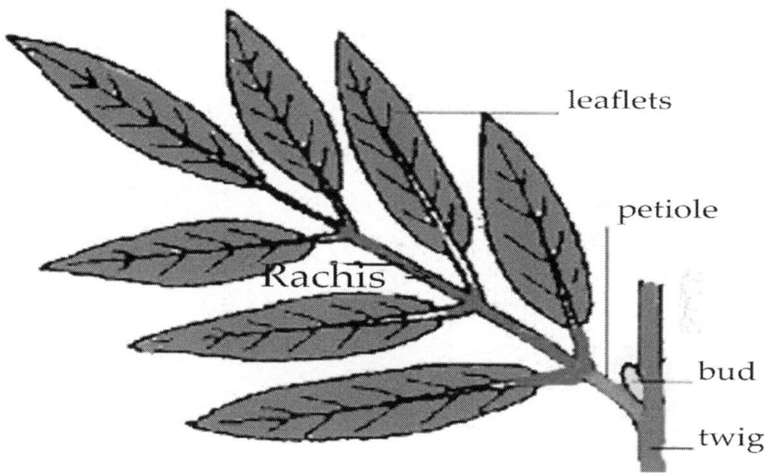

Figure 153: Rachis is an axis.

Source: image by Kapsenberg Website Design

Radicle

Radicle is a flowering annual plant that produces striking purple and white flowers. Radicle grows to be 6 inches tall and 4-6 inches wide. It blooms from June to August and attracts butterflies when it flowers. The foliage of Radicle is olive-green or purple, depending on the varieties, and the leaves are covered in fine hair. Radicle translates its name from the Latin word for "root", a reference to their focus on growing from their roots. Their collection encompasses their unique perspective of gardening's past, present, and future while challenging what many consider to be today's conventional wisdom. In most instances, radicle is used in referring to a plant's root or the first visible part of a seedling. In gardening and horticulture, it is the first developmental stage of a plant after germination and is analogous to the fetus in animals. During this stage, the stem of the plant is developed. The radical root absorbs water and minerals directly from the soil to help maintain healthy, nutrient-rich plants. The radical root extends the plant's reach, giving it a larger footprint in the ground while aerating the soil. It may be bare, or covered with soil particles or a mucilaginous sheath.

Rainforest

Figure 154: Rainforest.

Source: Picture by Rainforest Alliance

Rainforests are the world's oldest and most complex biomes. They have a wide variety of plant species, including many trees. They capture water that falls as rain, and release it through evaporation rather than allowing it to flow into rivers and lakes. The ecological functions of rainforests have made them among the most threatened ecosystems in the world. Many different kinds of plants and animals live in rain forests, including many kinds of birds. The world's rain forests are found in South America, Africa, Australia and Southeast Asia; some of the most extensive is found in Borneo, Brazil, the Congo Basin and Madagascar. The Amazon is one of the world's largest and most biodiverse rainforests (Waraich et al., 2011). Within the tropical rainforest biome, a typical region begins with a short dry season, followed by a prolonged wet season. Worryingly for plants, this climate results in a thick layer of vegetation that completely covers the forest floor and fills every available niche. Each niche has unique conditions to which plants must adapt. There are many thousands of species on this vast forested continent Tropical botany records reach from Southern Mexico all the way to northern Argentina, including the Amazon basin. In the Amazon itself, there are over 400 billion individual trees comprising 1500 species of tropical trees spread across approximately 7 million km2, or 2 million square miles.

Ramicaul

An exotic flower grown in tropical rain forests across the world. Ramicaul is a plant, with a flower so beautiful it will stop you in your tracks and make you stare. And like many other, more traditional fresh cut flowers, it only lasts a few days. Given its malleable nature, the flower petals and stem can be shaped into a variety of different shapes. It is also used as a decorative supply for crafts. The flower is used in religious ceremonies in China. Ramicaul combines a unique formula that delivers the active ingredients of Caulerpa Jackassiae, a seaweed derivative, and Guava leaf extract. With these exclusive plant extracts, Ramicaul is able to provide superior wound healing properties which complement its natural antioxidant action. The resulting clear gel dries quickly when applied. Its odorless and transparent nature makes it virtually invisible when applied to the skin's surface. Ramicaul is a flower extract that helps promote hair growth, slow down hair loss caused by telogen effluvium, and increases the viability of cells. Ramicaul is the standardized ethanolic extract obtained from the bulb of a selected Amaryllidaceae species (*Hyacinthoides hispanica* L.), a new efficiency-enhancing cosmetic active which can be used in different types of preparations: anti-hair loss shampoos, anti-hair loss foams, anti-hair loss serums or anti-hair loss spray dosed at 2% on the label of 100gms. Products successfully tested and commercialized in Spain (Solancept).

Figure 155: Ramicaul flowers are used in ceremonies.

Source: Picture by The American Orchid Society

Receptacle

Receptacle plants are grown using a patented semi-closed substrate system. This process provides highly concentrated substrates, conserves water and minimizes the amount of nutrients required to grow plants this size. The result is a more sustainable method of hydroponics by mitigating the amount of waste produced (Van Dijk et al., 2021).

Replum

Replum. A term applied in anatomy to the frame-like placenta of a capsule or other dehiscent fruit, across which the septum stretches. In taxonomy, it means a placenta-like structure that separates the true fruit from its covering. It is also used in the classification of flowering plants. It may refer to replum or femur, a hardened tip on the fruit capsule of some plants. The replum of "Carrot" is a strong and woody oval plate inclined to disappearance in the fruit of many Cruciferae. It exists very little in some of the Pigweeds, where instead the herbaceous part becomes flattened and embossed, with vascular bundles deeply sunken in this tissue. In Leguminosae, it is so tough as to furnish material for the arrow-head that decorates the native arrowhead. Its achenes in these genera are strung on flaxen thread, but their streamlets are external to the awns. It is a structure in angiosperms formed by the junction of a pistil and the top of the carpel, or to an inter-capillary region of tissue.

It develops into a fruit wall in syncarpous gynoecium but rarely occurs in gymnosperms. In contrast, it develops into the endosperm responsible for nourishing the growing embryo in seed plants without ovules.

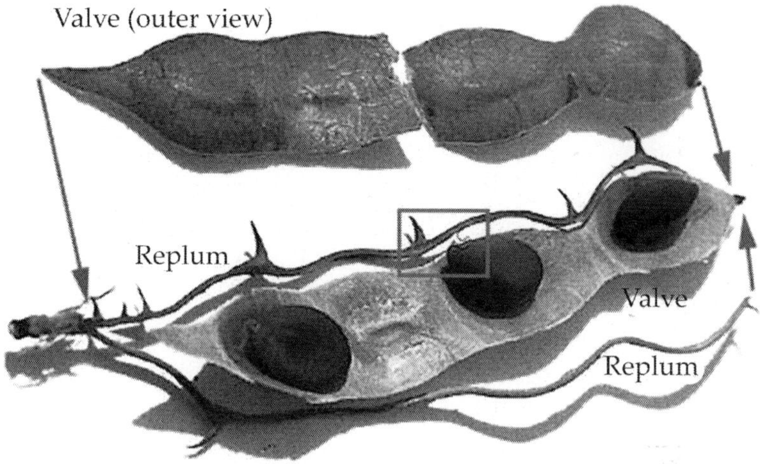

Figure 156: Replum on a mimosa pod.

Source: Picture by w3.biosci.utexas.edu

All the cells of the placenta are in uniform continuity with one another, but are completely separated from the solid tissue of the capsule. In many capsules, they all perish as they mature and leave only vesicles filled with fluid in a mass of undifferentiated tissue, but in others, the membranes remain distinct and become replum, septum, valves or teeth, which give way at maturity for the escape of the fluid contents.

Resupinate

Resupinate is a redefined word, which means twisting of flowers or leaves. The development of structures in certain flowering plants allows them to change the orientation of their petals or leaves from bud-scale to bloom. Resupination is a characteristic of the rosid subfamily Apiales, and specializes in angiosperms like carnations and petunias. For the case of flowers that face upward to encourage pollination, but also twist around to face down if they need water, this gives the plant two chances at getting pollinated (Van Bruggen et al., 2016). The resupinate flower arranges the dorsal (upper) side of its inflorescence at the base and its ventral (under) side at the apex. Putting these two views together, the plant presents its front to visitors from the ground, concealing the spiny or hairy underside of its leaves from potential herbivores. This is different from many other understorey plants that tend to present a uniform appearance from all angles as protection from above by predators.

Figure 157: Resupinate flowers.

Source: Picture by In Defense of Plants

Reticulate

The word "reticulate" (from the Latin for net) means covered with a network of lines or markings. When used in botany, it describes forms where the veins of leaves and flowers are arranged in a network or net-like pattern. Reticulate leaves and flowers form a distinctive geometric pattern. The reticulations are organized in a grid or net-like fashion, appearing on both the front and back of each leaf or flower. Reticulation is a pattern of branching veins or capillaries in a leaf. In botany, the reticulations in the leaves of certain plants such as *Cannabis sativa* are significant taxonomically. Reticulated wood is wood that is visible on cut surfaces of logs in which the internodal parenchyma has been lost, resulting in the formation of a three-dimensional network of xylem elements and phloem interconnecting strands. Reticulate leaf veins are a distinctive characteristic of a newly discovered palm (Van Breusegem et al., 2001). The leaves are green and approximately 6' in length with a petiole (stalk) that is long and red. Fruits and seeds are unknown. In addition, reticulate foliage is a feature of this remarkable tall shrub from South Africa. A plant that is covered with vivid red flowers towards flowering time, it will grace your garden for a good 10-15 years. This little beauty was planted about 3 years ago and is now 2.1m high by 1.5m wide.

Rhizine

Figure 158: Rhizines in Lichen.

Source: Picture by Wikipedia

Rhizine (or Rhizina) is a root-like structure in lichens, consisting of growth from the holdfast that has grown in the direction of propagation of the thallus. Rhizines originate from cells of the parent fungus and, in some species, they may act as ‘nodal lines’ of root-like growths. The rhizines are often important in attachment to surfaces, where they can also function in photosynthesis and absorption of nutrients. It serves to attach the lichen to its substrate through a combination of chemical and mechanical adaptations. In other words, it is what lichens anchor themselves to as they grow (Van der Kooi & Ollerton, 2020). Rhizine can be either branched or unbranched. Branched rhizines are attached directly to the thallus, while unbranched rhizines are attached by way of the basal disc, the upper surface of which may be tubular or supported by papillae shaped like inverted commas. Rhizine morphology varies among different types of lichens and also among different species of lichen. These root-like structures often take on a variety of forms, including unbranched, forked, clubbed, circular, or tubular. They are ubiquitous in the Bryophyta. The rhizoid is the non-reproductive part of any plant, and these roots take the place of the reproductive parts in plants. These roots can also be called rhizoids or rhizoids non-reproductive.

Rhizodermis

Among the primary layers that are found in roots, the outermost layer, called rhizodermis, is made up of polyhedral cells. The plant rhizodermis is a layer of root epidermis that lies outside the root proper. It is formed by the upward continuation of the underlying cortex and consists primarily of elongated cells, which are joined end-to-end with intercellular gaps. The rhizodermis gives rise to root hairs, the major absorptive and respiratory structures of the plant. It consists of four sub-layers: the endodermis, Casparian strip, Columella and pericycle. The casparian strip was described in 1827 by the German botanist Karl Ludwig Willdenow as "a small cartilaginous ridge or sheath in the epidermis of carnivorous plants, which prevents the loss of water from the aerial parts by evaporation." Casparian strips are quite common in primitive plants such as mosses. Columella is a spiral-shaped, plant support. It is typically a single leaf that extends from the root of a plant and down into the substrate providing stability for climbing plants or those that need additional support. Within the rhizodermis region, there is pericycle, consisting of branched cells in the form of a network and allowing free movement of water and air, continuous with endodermal phloem parenchyma.

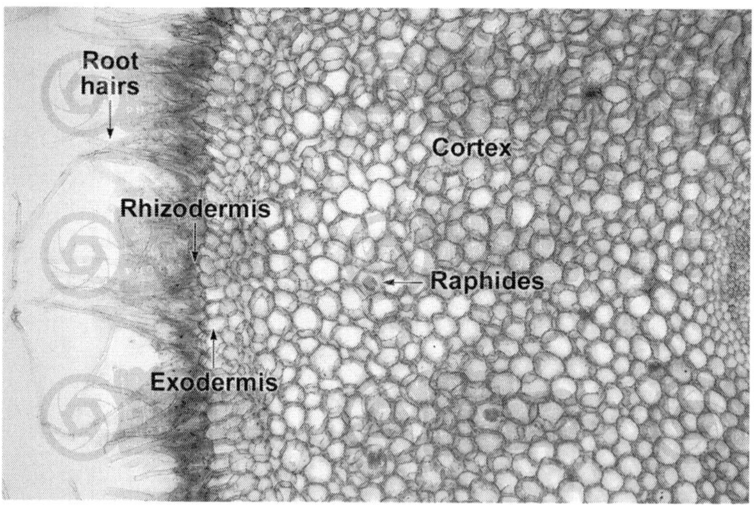

Figure 159: Rhizodermis.

Source: https://www.nature-microscope-photo-video.com/en/photos/plant-histology/plant-tissue/tessuti-tegumentali/rhizodermis.html

Rhizome

At this time the word rhizome refers to a modified subterranean plant stem that sends out roots and shoots from its nodes. A subterranean stem is also called a rootstock or rhizome, for example, ginger's rhizome is used as food and peppermint's rootstock is used for cooking. It can be eaten plain or with vinegar, syrup, or other condiments. Stems may be short and simple, such as yeasts and ferns, or long, often fleshy, like asparagus, bamboos and the common bean. Rhizome growth occurs from active nodes and not from terminal buds which are only involved in vegetative growth (Twidwell et al., 2013). These active nodes are usually internodal. The rhizome system is highly developed in palms and the lily family. The rhizome produces adventitious roots on the soil surface or in crevices of rock where they can receive minerals and water. Unlike true roots, rhizomes are horizontal. They grow horizontally under the soil and send up new stems or offshoots at the nodes, these in turn produce new rhizomes, etc. It is a characteristic of the true things (and is not found in the ferns).

Rhizomatous

A rhizomatous plant is a perennial plant (meaning it will re-grow from its roots year after year) that sprouts new plants through the growth of rhizomes, or horizontally running underground shoots. Rhizomatous plants are non-climbers, but they may produce aerial stems which can be either herbaceous (non-woody) or woody types. Many rhizomatous plants are used as ornamental plants. It is also a herbaceous perennial of moist ground, bearing egg-shaped or elongated leaves, with a slender stem bearing a spike or umbel of greenish-white flowers. The rootstock is horizontal and creeping; the aerial stems are erect, simple, branched, or panicled.

Figure 160: Rhizomatous.

Source: Picture by iStock

Rhizosphere

The rhizosphere is the layer of soil (rhizome) in which the roots of a plant grow. It can be thought of as the "interface between a plant and its environment", encompassing all of the chemical, physical and biological interactions between a plant and its environment. The rhizosphere is the specific area of soil that surrounds a root system. The rhizosphere can be found in any soil and contains larger amounts of nutrients than any other area of soil. It contains many different microorganisms, both useful and harmful to plants. The microbes living in the rhizosphere are responsible for breaking down organic matter and producing some nutrients for the plant to use (Taiz et al., 2015).

Rhytidome

The rhytidome features an elegant binding of plant and flower varieties selected for their evergreen and deciduous qualities. Succulents and cactus articulate the vertical borders while grasses, sedges, and ornamental grasses mingle with a strong foundation of Moor Grass, Yarrow and Blue Fescue. Delicate ferns harmonize beautifully with three kinds of pinks, Sweet Woodruff, and Eriocephalus. This biological structure functions in protecting vascular bundles, which carry the plant's vascular transport system that is responsible for food production, respiration, and growth. It also describes the plant tissues that are present outside the phellogen. Ideally, it is within and not separate from the plant.

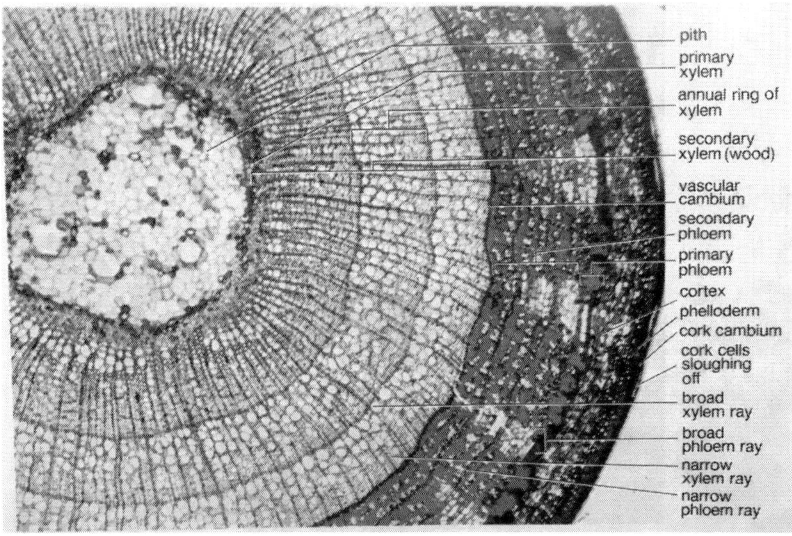

Figure 161: A plant Rhytidome.

Source: Picture by Quizlet

Root

The root is the plant underground part. It is made up of many cells, each of which is a potential new root. The root of a plant develops from a plant's cells. It absorbs water and food, and stores starches, sugars and fats. When plants die, the roots remain in the ground where they break down over time to form soil (Szpak, 2014). The root system is one of the 2 main parts of plant tissues. These tissues include both epidermises, or skin, and the cortex, which includes phloem (food transport tissue) and xylem (water transport tissue). Roots can also be aerial or aerating, that is, growing up above the ground or especially above water. The underground part of the root is called the rootstock.

Root hairs

This is the freely growing tip of the root of a plant or fungus, involved in the absorption of water and minerals. Root hairs are fine, threadlike organs found on the tips of plant roots. They absorb moisture and minerals from the soil, and are an important part of the plant's intake system. These specialized filaments are typically 1-5 mm long and allow for absorption of water and solutes (nutrients) from the soil and other sources. Plant roots have root hairs, which makes the connection between the plant and the earth. This helps the plant to absorb nutrients and water. Root hairs are very small with a diameter of 2 to 30 micrometers, so they can only hold this function because of their high surface area. The end of the root hair is pointed and surrounded by a horseshoe mesh called Casparian Strip; this part also helps make good contact with the soil. Root hairs act like a net, but once plucked out from the soil, it stops functioning as one.

Figure 162: Root hairs on plant roots.

Source: Picture by Farmtario

Root microbiome

The microbial community associated with the root microbiome varies considerably depending on plant species, environmental and geographic location, and soil type. Characterizing the structure and function of plant root microbiomes is key to understanding how plants access resources from soils and respond to changes in the environment (Stenberg, 2017). The root microbiome refers to the interaction between bacteria or other microbes and the plants to which they are attached. The microbiome of a rice plant, for instance, is distinct from that of a wheat plant although both plants are of the same family (Poaceae).

Rootstock

The part of a plant that is below the ground and is attached to the stem. It has buds, the main root (which grows from the bud), and other roots that bring minerals, water, and food to the plant. Roots grow out from the bud that forms at the tip of a rootstock. They grow downward and give support to the leaves and stems. The rootstock of a young grapevine is the part on which buds grow, creating baby vines and providing an additional source of food for the plant.

Rosette

Figure 163: Rosette on a young plant.

Source: Picture by NC State Extension Publications

A rosette, often referred to as a bud or baby crown, can be removed from a plant while it is still young and small to propagate a new plant. Simply take a cutting of the plant and replant it in the soil. They can be described as leaf-like structures at the end of some flower stems like daffodils, roses, and hibiscus. The petals of these flowers swirl gorgeously around a central disk in a sunburst shape. The disk contains the flower's reproductive parts and is usually very colorful.

Ruderal

Ruderal parts are the plant parts that are grown in a cultivated area, as opposed to those growing on wild uncultivated land. The term is used to cover any wild-collected and forest-derived material. Plants in the genus ruderal grow close to the ground, and are characterized by rapid growth, early maturation and short stature (Singh, 2015).

Rudiment

The rudiment is a small organ present in every plant. It sometimes is also called the rudimentary organ because it retains some function of the original embryonic tissue. It is a plant structure found in stems, roots, and leaves of plants. It is described as a factor that develops a convergent way to carry out various processes for the plant such as Xylem and Sporangium.

S

Senescence

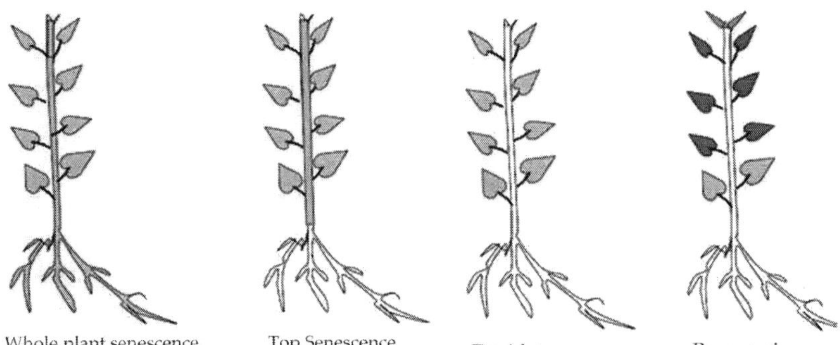

Figure 164: Senescence is the final stage of plant growth.

Source: https://qforquestions.com/what-is-senescence/

Senescence is the last phase of plant development whereby the plant recovers the significant cell building blocks which have been placed in the leaves and different sections of the plant during development. Having a proficient senescence process is fundamental for the durability of the plant or its future propagation. Senescence is a complex profoundly regulated process that requires fresh gene articulation and includes the associations of many flagging pathways.

Senescence in crops is a multifaceted, formatively managed stage during which cell structures are painstakingly destroyed and the dissolvable parts moved to different sections of the plant, either for capacity or to advance further development (Schuurink et al., 2006). All sections of the plant go through senescence however with the end goal of this, the accentuation will principally be on the senescence of leaves since this has been the most examined lately. Leaf senescence happens when the leaf is at the point of not utilizing the plant. This might be because of the progressive phase of the plant, the age of the leaf or be stimulated by environmental elements.

As plants enter the blossoming and seed advancement stage, the supplements inside the leaves are needed for the seed improvement thus, for this situation, senescence is initiated by formative signs. On the other hand, plant development makes the leaves at the foundation of the plant to become concealed and limited in their photosynthetic potential, for this case senescence is instigated by ecologically controlled signs. Different kinds of natural pressure, for example, water and nutrient constraint or microbe infiltration can likewise influence plant development and achieve premature senescence. Senescence is at its generally obvious in the pre-winter when the sensational shades of deciduous trees are striking tokens of the evolving seasons. This senescence is brought about by occasional changes like light and temperature.

Stomata

Tiny openings, lined by guard cells in the leaf epidermis and stems whereby gases pass (CO2, O2, H2O-fume); the whole stomatal mechanical system (guard cells in addition to pores). Guard Cell: Set of specific epidermal cells encompassing a pore, or stoma; variations in turgor tension of several guard cells cause opening and shutting of the pore.

Stomata are little openings or pores in plant tissue performing gas exchange. Stomata are normally found in plant leaves however can likewise be found in certain stems. Particular cells known as guard cells encompass stomata and work to open and close stomatal pores. Stomata permit a plant to take in carbon dioxide, which is required for photosynthesis. They additionally help to decrease water loss by shutting when conditions are hot or dry. Stomata resemble little mouths which open and close as necessary (Smith & Knapp, 2001).

Terrestrial plants commonly have a great many stomata across their leaf surface. Most stomata are situated on the bottom side of plant leaves decreasing their exposure to temperature and air current. In amphibian plants, stomata are situated on the upper surface of the leaves. A stoma is encircled by two kinds of unique plant cells that contrast from other plant epidermal cells. They are known as auxiliary and guard cells.

Stomatal Regulation

Stomatal movements arise from variations in turgor force around the guard cells. The significant solute liable for this inclination is potassium (K^+); greater K^+ and Cl-fixation causes stomata to release (water surges in because of absorption), though closes when it falls. Opening happens when solutes are effectively gathered in these cells. Stomatal opening is achieved by the opposite interaction (a reducing monitor cell solute); water moves out of the guard cells bringing down turgor pressure. The guard cells chloroplasts process CO_2 photosynthetically to create sugar, which adds to the solute development needed for stomatal opening.

Sage

Sage or Salvia officinalis is a perpetual, evergreen subshrub, with wooden stems, greyish leaves, and azure to purplish blossoms. It forms part of the mint class Lamiaceae and is local to the Mediterranean area, however it has been naturalized in many spots all through the globe. Sage is taken for stomach-related issues, such as appetite loss, gas (tooting), stomach ache (gastritis), bulging, and acid reflux. It is additionally utilized for diminishing overproduction of sweat and saliva; and for depression, cognitive decline, and Alzheimer's illness (Simon & Petrášek, 2011). Under unique handling, the crop can be harvested twice or thrice each year (from the second year forwards), making it a beneficial specialty plant, assuming that the local conditions are ideal. The crop is exceptionally strong and versatile and can flourish even in malnourished soil.

Safety Margin

Leaf pressure safety margin (HSM_{leaf}) is determined as the distinction between least water potential under normal drought circumstances (Ψ_{min}) and $P50_{leaf}$. Restricted or negative HSM_{leaf} demonstrates a great probability of leaf hydraulic risk under dry season. The vulnerability division theory recommends that leaves are more defenseless against cavitation than stems, and an early decline in leaf pressure conductance could forestall water loss and shield the primary stems from water-based breaking. To be sure, species with smaller HSMleaf figures will generally have a more grounded leaf-to-stem weakness lines. Besides, a new report from occasionally dry tropical woods reports a huge connection between HSMleaf and death rates during extreme dry seasons, demonstrating that HSMleaf might assume a part in foreseeing tree demographic rates under environmental change.

Scouting fields/Crops

Figure 165: A farmer scouting his vegetable field .

Source: https://www.growingproduce.com/vegetables/when-in-doubt-about-your-vegetable-crop-scout-it-out/

Crop scouting, otherwise called field scouting, is the primary activity of going through a harvest field, typically by walking, while at the same time mentioning regular stops for objective facts. Crop scouting is done as such that a rancher can distinguish how various spaces of their field are developing and what risks or bugs might be available. Assuming there are issues during the developmental stage, the rancher can attempt to moderate them so those issues don't influence yield during harvest time.

The recurrence with which fields are scouted relies upon the kind of plant grown, phase of yield life cycle, current and anticipated pest(s) and their lifecycle, and natural conditions (climate). In light of these factors, field visits ought to be arranged to guarantee that yield sets up a good foundation for itself, that development continues with the solid turn of events and doesn't continue into a populace bottleneck, and that vermin don't cross the economic limit level, later which could turn into a significant issue (Rusman et al., 2019). This is for the most part not an issue when the yield has arrived at a particular stage of development, so the amount of visits can be diminished. Recurrence of visits to the fields ought not to be diminished, and these fields ought to be completely checked.

Seed Treatments

Pesticides applied instantly to a seed prior to planting, to ensure seeds, seedlings, and crops from pests. Seed treatment means the use of fungicide, insecticide, or a mix of both, to seeds in order to sanitize and clean them from seed-borne or soil-borne pathogenic organisms and insects. It additionally means the repressing of seeds to sun-based energy exposure, inundation in treated water, and so forth. The seed treatment is done to accomplish the associated benefits. Advantages of Seed Treatment: 1) Avoids spread of plant illnesses 2) Guards seed from seed decay and plant blight 3) Enhances germination 4) Provides security from storage bugs 5) Controls soil bugs.

Forms of Seed Treatment:

a) Seed sterilization: Seed sanitization means the destruction of parasitic spores that have become lodged inside the seed coat, or all the more well-established tissues. For successful control, the fungicidal treatment should really enter the seed to kill the growth that is available. b) Seed disinfestation: Seed disinfestation alludes to the obliteration of surface-borne organisms that have degraded the seed surface but not tainted the seed surface. Compound dips, drenches, fungicides applied as residue, or fluids have been seen as effective. c) Seed Protection: The reason for seed protection is to shield the seed and young seedlings from organisms in the soil which may somehow cause rot of the seed before germination.

Conditions under which seed should be disinfected:

a) Injured Seeds: Any break in the seed layer of a seed determines the risk for parasites to enter the seed and either kill it, or stir the seedling that will be created from it. Seeds experience mechanical injury during sorting and sifting activities, or from being dropped from unreasonable heights (Reid et al., 2009). They may likewise be harmed by climate or inappropriate stockpiling. b) Diseased seed: Seed might be tainted by illness-causing organisms at the period of harvest, or may become contaminated during handling, whenever handled on dirty apparatus or whenever stored in polluted holders or stockrooms. c) Undesirable soil conditions; Seeds are occasionally planted under negative soil conditions like cold and sodden soils, or incredibly dry soils. Such undesirable soil conditions might be good for the development and improvement of specific parasitic spores allowing them to harm the seeds. 4) Illness-free seed: Seeds are constantly infected, by illness microorganisms that may cause decay. Seed treatment gives decent protection against sicknesses, soil-borne pathogens and in this manner allows them to sprout and create seedlings.

Saponaria

Saponaria is a variety of blossoming plants in the Caryophyllaceae family, local to Asia and Europe, usually referred to as soapworts. These are herbaceous annuals and perennials, some with wooden bases. The blossoms are plentiful, five-petalled and generally in colors pink or white. The class is connected to Lychnis and Silene, being recognized from these by having just two (not three or five) styles in the blossom. It is likewise connected with Gypsophila, however, its calyx is round and hollow as opposed to bell-formed. The most natural species may be natural soapwort (*S. officinalis*), which is local to Eurasia however is referred to in a large part of the world as a common weed, and some of the time an ornamental plant (Poff & Zimmerman, 2010). The family name Saponaria comes from the Latin sapo ("cleanser") and - aria ("relating to"), and has been commonly utilized to make cleansers. It contains saponins, and a fluid cleanser could be delivered by absorbing the leaf's water. This cleanser is commonly used to clean fragile classical woven artworks. Saponaria species are consumed by the hatchlings of certain butterflies and moths, including the Lychnis and *Coleophora saponariella*, which is unique to the class.

Soil Organic Matter

Soil Organic matter (SOM) is whatever was once alive and is currently in or on the soil as it is decayed into humus. Humus is a natural matter that has been deteriorated by microorganisms and is promptly changing structure and mass as it breaks down. Soil natural matter is the natural part of soil, comprising of three essential parts including small plant growths and living soil organisms, decaying matter, and humus. Soil decay fills in as a supply of supplements for crops, holds moisture, lessens compaction, decreases surface crusting, and builds water infiltration into the soil.

Plant build-ups on the soil surface like leaves, fertilizer, or yield build-up are not considered SOM and are normally eliminated from soil tests by sieving across a 2 mm wire network before review. Soil natural matter can be assessed in the field and tried in a lab to give evaluations to Nitrogen, Phosphorus and Sulfur mineralized available for crop production.

Soil organic material affects the rate of surface-applied herbicides alongside soil pH important to viably control weeds. Soil organic material affects the potential for herbicide deposit for future harvests, and quantity of lime needed to increase pH.

Strip-till

Strip till is a field cultivation system that consolidates no tillage and full culturing to create column crops. Slender strips 6 to 12 inches wide are plowed in crop stubble, with the space between the columns left undisturbed. Regularly, manure is infused into the ploughed region during the strip-plowing activity. "Strip-till" is a process where buildup free portions of soil are plowed in rather than utilizing a blade contraption, for example, a manure infusion knife (Pauchard et al., 2009). The strips are roughly 6 inches wide or around 1/3 the column width and 4 to 8 inches down. These strips are free from build-up and plowed for warming and drying purposes either previously or during the planting process. Manure is frequently fused periodically and strip culturing is done for better seed position. The seeds are planted straightforwardly into the piece of tilled soil.

Stonecrops

Stonecrops might be the most unique group of succulents, with diverse varieties that vary from surface covers to bushes. They're among the toughest succulents as well. Stonecrops are an assorted group, so pick the one that is best for your flower nursery (and afterward one year from now, you can add another "best.") Most sedums are solid in tepid environments, making creeping types dependable ground covers. Others, similar to Autumn Joy', have a place in a perennial site where their late-season blossoms will bring trust even to a groundskeeper who realizes winter is ahead. Creased-leafed *Sedum cauticola* comes from Japan and has gleaming green foliage that embraces the ground. *Sedum spurium* is an evergreen and its profound purple leaves will make a brilliant carpet.

Sea Lavender

Sea lavender is principally grown for its delightful botanical display; however its enormous green oar style basal rosette leaves will likewise make an appealing ground-cover in the late-winter, until the plant blossoms in the late spring. Some part of what makes sea lavender perpetual is that the leaves are so near the ground, and long blossom stems will create from this gathering of foliage. Sea lavender is a genuinely easy perennial to grow and has not very many requirements. It tends to be isolated every few years (consistently in late-winter), however make certain to burrow well to forestall damage to its long roots (Prajapati & Modi, 2012). Limonium is a class of 120 blooming plant species. Individuals are otherwise called sea lavender, caspia or statice. In spite of their typical names, the species are not connected with the lavenders or to rosemary. They are rather in the Plumbaginaceae, the plumbago and leadwort family.

Saprophyte

Plant consuming decayed foliage in the soil. A saprophyte is a crop that doesn't have chlorophyll and derives its food from dead material, like microscopic organisms and growths (fungi are regularly called saprophytes, which is wrong, since parasites are not plants). These plants utilize compounds to change over natural food materials into simpler structures from which they can ingest supplements (Figure 2). Most saprophytes don't instantly process dead matter: all things considered, they parasitize growths that digest dead matter, or are mycorrhizal, eventually deriving photosynthate from an organism that got photosynthate from their host. Saprophytic crops are remarkable; a couple of plant groups are defined.

Symbionts

A symbiont is a crop in a cooperative relationship, with unique variations, for example, mycorrhizae or knob formation. Parasites additionally form cooperative relationships with cyanobacteria and green growth (called lichens). Lichens can in some cases be viewed as clear developments on the outer layer of rocks and trees. The algal matter (phycobiont) produces food autotrophically, some of which it imparts to the parasite; the fungus (mycobiont) retains water and minerals from the climate, which are made accessible to the green alga. In case one partner was isolated from the other, the two of them would deteriorate away (Neumann et al., 2009).

Scape

Leafless stem bearing blossoms. Leafless unbranched blossom tail with one bloom developing straight from the ground surface in a tulip, peduncle. Inplant science, scapes are blooming stems (peduncles), normally leafless, ascending from the crown or foundations of a plant or from a bulb or basal rosette at ground level with one blossom. All things considered, the term scape is applied to demonstrate any stem which upholds the blossoms, buds, and seed units, and show a wide number of very unique blooming stem designs and inflorescences, a few scapes have scales, foliaceous bracts and sometimes even branches, and some scapes can have a single bloom or many blossoms, based upon the species.

Secund

Looking a similar way. A botanical or foliar organ is called secund when each of its parts is branched aside. A branchlet is secund when every one of its leaves is viewed as on one side and not on both. You may portray a leaf as secund when the lamina on one or the other side of the rachis is arranged above or underneath the pivot (V-shaped). The conflorescence is secund when every one of the blossoms are on one section of the node, toothbrushed inflorescences for instance - those of *Grevillea rigida*.

Sepal

External line or ring of perianth sections developing the defensive cover of a bloom bud.

A sepal is a piece of the blossom of angiosperms (blooming plants). Generally green, sepals typically work as security for the blossom in bud, and frequently as backing for the petals when in sprout. The word sepalum was instituted in 1790 by Noël Martin Joseph de Necker, and derived from the Greek term 'covering'. On the whole, the sepals are known as the calyx (plural calyces), the furthest whorl of parts that form a bloom. The term calyx was derived from the Latin calyx, not to be mistaken for calix 'cup or goblet'.

Subsequent to blooming, most plants have no more need for the calyx which wilts or becomes minimal. A few plants hold a prickly calyx, either dried or live, as insurance for the natural product or seeds. Examples incorporate types of Acaena, a portion of the Solanaceae (for instance the Tomatillo, *Physalis philadelphica*), including the water caltrop, *Trapa natans*. In certain species the calyx exists in the wake of blossoming but rather than wilting, starts to develop until it develops a bladder-like nook around the fruit product (Nikunj et al., 2014). This is successful security against certain sorts of birds and bugs, for instance in *Hibiscus trionum* as well as the Cape gooseberry. In different species, the calyx develops into an extra fruit product. Morphologically, the two sepals and petals are transformed leaves. The calyx (the sepals) and the corolla (the petals) are the external sterile whorls of the blossom, which together constitute what is called the perianth. Similar to traditional leaves, sepals are fit for performing photosynthesis. Notwithstanding, photosynthesis in sepals happens at a more slow rate than in traditional leaves because of sepals have a lower stomatal thickness which restricts the holes for gas trading.

Septum

A slender segment or film dividing single seeds inside a seed case or fruit product. In crops, as in organisms, the septum is a segment that divides in at least two chambers, known as locules. These are present in anthers (the dust-bearing piece of a stamen), fruit products, and sporangiums (found on spore-forming plants, like ferns). The word "septum" is utilized in both animal and plant science to allude to a design that divides at least two chambers. In plants, maybe the best illustration of a septum would be the design inside some fruit products that isolates the seeds into various cavities, or locules.

A few fruit products lack septa totally, whereas others might have at least one. Consider an orange – when you eliminate the peel, the inside of the fruit product is an assortment of independent segments, what is considered as cuts. Each cut is unique, and comprises of its unique seeds, just as tissue. One more illustration of a septum in the realm of organic products is the tomato (Novello et al., 2020). When you cut a tomato in half, you'll find that there is a firm septum partitioning the inside into particular chambers, each loaded up with seeds. The tomato septum is the "meat", and various assortments have more slender or thicker septa. Beefsteak tomato types have not very many seeds, and an exceptionally enormous septum. Cherry tomatoes, contrarily, have next to no meat, and more seeds.

Sessile

Without a stem. Plants are sessile living beings that are forever confined to their point of germination. To make up for their absence of movement, plants evolved extraordinary tools empowering them to quickly respond to changing ecological conditions and deftly adjust their postembryonic formative development. A conspicuous exhibition of this formative pliancy is their capacity to twist organs to arrive at the position generally ideal for the development and use of light, supplements, and different resources. Soon after germination, dicotyledonous seedlings structure a twisted construction, the purported apical snare, to secure the sensitive shoot meristem and cotyledons develop a bent posture while entering through the soil (Mercuri et al., 2002). With a continuous supply of light, the apical snare quickly opens and the photomorphogenic formative program is instituted. During later germination, plant organs can adjust their development to the light source and embrace the most positive direction through bowing, in a cycle named phototropism. Then again, when roots and shoots are redirected from their upstanding direction, they promptly recognize an adjustment of the gravity vector and adapt to keep an upward development course. Notwithstanding, all plant developments share a common differential cell development process.

Silicula

The silique or siliqua is a form of natural fruit (seed capsule) containing two combined carpels having a length that is twice or thrice the width. When the length is below thrice the dried fruit's width, the plant is called a silicle. The external ovarian walls (the valves) generally separate when ready,at that point called dehiscent, and then leave a determined segment (the replum). Siliques are available in multiple individual forms from the mustard family, Brassicaceae, yet a few plant categories have silicles all things being equal. A few plant varieties with genuine siliques have fruit products with a comparative design that don't open when ready; these are generally known as indehiscent siliques (contrast dehiscence).

Fruit product of the cabbage family, regularly rounded and rather longer than it is wide. Silicula is a kind of capsular dehiscent natural fruit with massive, syncarpous ovary having a bogus septum known as replum. It is a flat and leveled form of siliqua fruit product. Achene and schizocarpic consists of two other forms of natural fruits (Matyjaszczyk & Śmiechowska, 2019).

Spathe

Enormous hooded bract encasing a spadix. The spathe is a solitary bract that encompasses the spadix, which is a blooming spike. It is ordinarily dense and beefy, having extremely small blossoms bunched on it. A reality about the spadix is that in certain plants it really generates heat, presumably to draw in pollinators. The spathe resembles an enormous bloom petal, however, it is really a bract. A bract is a changed leaf and is regularly shaded and stands apart more than the real bloom does. Poinsettia is an illustration of a plant with flashy bracts. A spathe is a solitary bract that encompasses the spadix, which is a blossoming spike. It is regularly thick and plump, having exceptionally minuscule blossoms grouped on it. You will be unable to tell if these are really blossoms. A pleasant fact about the spadix is that in certain plants it really delivers heat, likely to draw in pollinators.

Species

The term plant species means the second piece of a crop's scientific name. Utilizing the analogy of an individual's name – their last name and forename – and relating it to the class name and species name frequently assists with clarifying the usage of these Latin plant names and makes it clearer to comprehend. Order of a class of comparative plants that breed valid in nature.

- Researchers have distinguished in excess of 260,000 types of plants.
- They group plants based on body parts like seeds, roots, stems, leaves, and tubes.
- The three fundamental groups of plants are ferns, mosses, and seed plants,

Speculum

Ophrys speculum, commonly called the mirror orchid, is a type of Ophrys dispersed all through the Mediterranean and pollinated only by a solitary type of scoliid wasp. A land based orchid up to 25 cm tall and every inflorescence delivers somewhere in the range of 2 and 8 huge blossoms. The plants frequently develop in groups. In bright daylight the blossoms are exceptionally noticeable as the light reflects off the speculum in the focal point of the lip – it is a brilliant glowing purple/blue in shading and extremely lustrous (Mandel, 2019. The lip is three-lobed and lined by a greenish-yellow edge which is encircled by a band of thick smooth hairs which are ruddy brown. The sepals and petals are green and separated by violet marks and stripes.

Spike

Ungrouped bunch of roses that are unstalked. A spike means a kind of inflorescence (blossom structure) that delivers higher-than-average blossoms, most as a rule on a pivot that is unbranched. Like a raceme, the spike bears blossoms that are directly appended to the plant, with no stalks at all. Spikelets are more modest spikes consisting of a few florets that are encased by sterile bracts.

Spikes are one form of inflorescence, which are bunches of roses that are ordinarily found in a complex structure of stems and branches. Not at all like raceme inflorescences, spike flowers lack pedicels. All things considered, they open directly from the base to the actual top of the plant (Lindström et al., 2013. One of the well-known complex spikes happens on Spiranthes. Having a place within the orchid family, this plant has frequently been contrasted with hair braids on account of its interlaced spikes. Common examples are wheat and rice.

Stamen

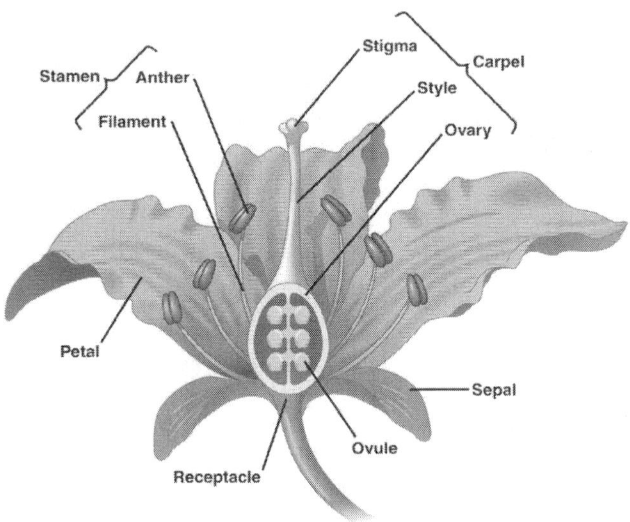

Figure 166: Stamen contains the male organs of a flower.

Source: https://socratic.org/questions/what-is-a-stamen

Male reproductive part of a blossom. A stamen is the male regenerative organ of a blossom. The stamen is comprised of 2 sections: the anther that is held up by a fiber. The anther produces dust grains that are utilized during fertilization. The dust is delivered when the bloom opens its anther (Luan et al., 2020. The blossom depends on organisms, bugs and the breeze to deliver the dust to another blossom. The pistil is the female section of a blossom. A blossom that has both male and female parts is known as an ideal blossom. A blossom with just male or female parts is a defective bloom.

Spur

Empty pocket, round and hollow or tapered, rising from a blossom and containing nectar. Spurs are thickset stems that project from the primary stem on woody plants that are ordinarily in their second year. The pear and apple fruit structures develop from the tree spurs. Since spurs are vital to fruit-bearing trees and plants, pruning procedures are ordinarily utilized that will allow for their development. In a grape plantation, grapes are regularly pruned down to the spike to expand natural fruit production. Blossoming trees can likewise be pruned to support prod growth and further improve flowering.

The botanical nectar spur is an empty extension of specific blossom parts that regularly comprises tissues delivering nectar. The nectar spur is the point of convergence of numerous transformative and environmental speculations, and it was believed by Darwin to see how natural selection improves species qualities (Lambers et al., 2008). It turned into a significant piece of the fertilization disorder conundrum and, later, the spur was viewed as a key development that caused the quick expansion of specific plant phylogenetic genealogies. Besides, the nectar spur assisted with understanding both individual plant–pollinator communications and the relations of plant–pollinator networks. As of late, the nectar spur has been viewed as a significant trait managing natural invasions.

Standard/Standard Petal

Upstanding, the upper petal of a pea blossom that is bigger than the others. Papilionaceous blossoms are blossoms with the trademark random and butterfly-like corolla found in many, however not all, plants of the Faboideae legume subfamily.

The blossoms have a two-sided evenness with the corolla comprising of five petals. A solitary, huge, upper petal is known as the pennant (additionally vexillum or standard petal). The semi-tubular-shaped base of the bunter packs two equivalent and more modest side wings (or alae). The wings thusly encase a couple of little fall petals, which are arranged to some degree lower than the wings, yet are inside them. They have inward sides and compare with the state of the wings. The two fall petals are combined at their bases or remain together to form a boat-molded design that encases the primary blossom organs, in particular the androecium and gynoecium. Normally these blossoms have a vexillary (for example descendingly imbricate or covering) aestivation (Klassen, 2001).

The stamens frequently have a diadelphous morphology, implying that they are joined in a pair with a unique fiber shape per set. The lesser set establishes a membranous sheath at its base, which encompasses the single, predominant pistil in a cylinder. The cylinder isolates into nine fibers, which follow the foundation of the fall petals that wall them in. The single free fiber lies above them. In Smithia the pair are equivalent, with 5 stamens in each. Numerous papilionaceous blossoms nevertheless have ten unique stamens. Genera that adapt to the last morphology include: Cercis, Daviesia, Dillwynia, Sophora, Anagyris Mirbelia, Pultenaea and, Gompholobium,

Stemless

A plant lacking a conspicuous stem however with a blossom tail emerging directly from the earliest stage. Stemless plants have no noticeable stem over the ground and are made mostly out of leaves or leaf like products. Having no evident stem or trunk over the ground. Of plant propensity, with no over the ground stem aside from an inflorescence node. A plant that comes up short on an aeronautical stem or having an exceptionally short stem or trunk which doesn't show up over the ground, however, there might be an underground stem (tuber, rhizome, and so on). Meaning plants that have no focal stem or just an extremely short stem so that, where present, every one of the leaves and inflorescence is bunched at or close to the foundation of the plant at ground level. The blossom might be on an erect peduncle yet this will have no genuine leaves. Examples consist of dandelion, Aloe vera, sisal, onion and liverworts.

Stigma

Part of a blossom that gets pollen dust. The stigma is an adhesive knob at the highest point of the pistil. It is appended to the elongated, tubelike construction called the style. The style passes onto the ovary that contains the female egg cells called ovules. The male parts are called stamens and ordinarily encompass the pistil. It very well might be fluffy and expanded or lengthened, as in such wind-pollinated blossoms as those of the grasses, or it could be conservative and have a tacky surface (Katova, 2007).

The stigma, along with the style and ovary includes the pistil, which is important for the gynoecium or female conceptive plant organ. The actual stigma shapes the distal piece of the style, or stylodia, and is made out of stigmatic papillae, the cells of which are open to dust. These might be confined to the summit of the style or, particularly in wind-pollinated plants, comprise a wide surface.

The stigma gets dust and it is on the stigma that the dust grain develops. Frequently tacky, the stigma is adjusted in different ways to catch and trap dust with different hairs, folds, or sculpturings. The dust might be caught from the air (wind-borne dust, anemophily), from visiting bugs or different organisms (biotic fertilization), or in uncommon cases from surrounding water (hydrophily). Stigma can vary from long and thin to globe formed to feathery. Dust is normally parched when it leaves an anther. They have been shown to aid the rehydration of dust and in promoting germination of the dust tube. Stigma likewise guarantees appropriate attachment of the right types of dust. Stigma can assume a functioning part in dust segregation and some self-incongruence responses, that reject dust from something similar or hereditarily comparable plants, include the connection between the stigma and the outer layer of the dust grain (Jeger et al., 2007).

The stigma is frequently parted into projections, for example trifid (three-lobed), and may look like the top of a pin (capitate), or come to a point (punctiform). The state of the stigma might shift considerably: Corn marks of stigme, by and large called "silk" come in different shapes and sizes.

Style Structure

The style is a restricted vertical augmentation of the ovary, interfacing it to the stigmatic papillae. Occasionally it could be missing, in which case the stigma is depicted as sessile. Styles are cylinder-like—either long or short. The style can be open (containing not many or no cells in the focal center) with a focal channel which might be filled up with adhesive. On the other hand, the style might be shut (thickly loaded with cells all through). Most syncarpous monocots and some eudicots have open styles, while numerous syncarpous eudicots and grasses have shut (strong) styles containing specific secretory transfer tissue, connecting the stigma to the focal point of the ovary. This forms a supplement-rich lot for dust tube development. Where there are more than one carpel to the pistil, each might have a different style-like stylodium, or offer a typical style. In Irises and others in the family Iridaceae, the style isolates into three petal-like (petaloid) style branches (known as 'stylodia'), nearly to the foundation of the style and is known as a tribrachiate. These are tissue folds, running from the perianth tube over the sepal (Hirschberg, 2001).

The stigma consists of an edge or rim on the bottom of the branch, close to the terminal lobes. Style branches likewise form on Dietes, Pardanthopsis and various types of Moraea. For Crocus there are three partitioned style branches, making a tube. Hesperantha forms a scattered style branch. On the other hand, the style might be lobed instead of fanned. Gladiolus comprises a two-lobed style branch (bilobate). Romulea, Savannosiphon, Watsonia, Freesia, and Lapeirousia have bifurcated (dual extended) and recurved design branches.

Stipule

Leaf-style organ at the bottom of a leaf tail. Stipules are the horizontal parts that are seen as joined at the foundation of the leaf. Stipules primarily secure the bud and arising young leaves. In certain plants, stipules perform photosynthesis. A few stipules become thistles, spines and shield the plant from destruction by predators. The spine is actually a transformed, sharp-pointed leaf. Because it has a bud within its axil, the spine forms in the general spot of the leaf. Some spines are classified "stipular spines" since they are transformed, sharp-pointed stipules at the foundation of a leaf.

They are for the most part situated close to the foundation of a leaf. Stipules are typically found on dicotyledons and in some monocotyledon plants have stipule-like constructions. There are various sorts of stipules present in plants like layered, intrapetiolar, interpetiolar, sidelong, ventral, ochreate, curled and spinous. The petiole is (plant science) the tail of a leaf, connecting the sharp edge to the stem while stipule is the basal end of the leaf of a blossoming plant (Hadwiger, 2013).

Stolon

Stem maturing on a level plane above or underneath the ground. A stolon is a thin stem that develops evenly along the ground, leading to roots and elevated (vertical) branches at particular spots called nodes.

a. Adventitious roots are formed at the nodes, and runner buds can form into upstanding shoots, and separate plants can be shaped, for example, strawberry plants.

b. A rhizome is like a stolon yet develops underneath the ground, for example, common potato.

c. A tuber is actually an adjusted stolon. The stolon closures swell to develop tubers. The tuber is subsequently an enlarged stem. The potato is a notable example with 'eyes' (buds) which develop shoots when the potato tuber is planted. Potato tubers will likewise become green by forming chlorophyll when exposed to light; this is one more characteristic of stems.

d. A tendril is present on specific legumes, like the pea. The peripheral leaflet is altered and is valuable for climbing reasons to sustain the plant.

Style

The piece of the female reproductive part that connects the ovary to stigma. Styles could be free or, all the more regularly, joined together, and they could be either lobate, with distinct stigmatic projections, or basic, which is the most well-known status in the Asparagales. Within the Iridaceae subfamily, the style is partitioned into three wide blades.

The blossoms in the class are likewise remarkably varied, going from the little, camouflaged, white-to-greenish, radially symmetric (actinomorphic) blossoms of most Asparagaceae to the huge, splendidly shaded blossoms of Amaryllidaceae, Iridaceae, and Orchidaceae. The corona, a petaloid expansion of a few or every one of the tepals and maybe generally clear as the trumpet sections of the blossoms of Narcissus, is present in some Amaryllidaceae (Guo & Liu, 2007).

In most Asparagales species, the blossoms are borne in terminal inflorescences on airborne stems that might carry ordinary or limited leaves; if leafless, the blossoming stem is regularly called a scape. Scapose inflorescences define a large number of the species that have bulbs and are common of those Asparagales found in the Amaryllidaceae group.

The aerial stem is abbreviated (decreased) in some Orchidaceae and Iridaceae plant species. Accordingly, the blossoms are developed at ground level, frequently with the ovary subterranean and at the foundation of a tubular flower. Among the notable stemless genera is the Crocus of

Iridaceae. A lot more models can be found in arid environments, like the Middle East and Southern Africa. While the ovary might be underground at blossoming, the blossom stalk (peduncle) typically extends, so the ovary is a brief distance over the ground as the seeds grow and age.

Subshrub

Tiny perennial with woody stems. A subshrub or bantam bush is a short woody plant. "Subshrub" is frequently utilized reciprocally with "bush". Because the measures are matters of extent rather than of kind, the meaning of a subshrub isn't directly recognizable from that of a bush; cases of reasons behind depicting plants as subshrubs incorporate ground-based stems or low development predisposition. Subshrubs may be generally herbaceous, with overwintering perpetual woody stems much lower-forming than deciduous summer growth. A few plants portrayed as subshrubs are just slightly woody and some continue for just a couple of years; others notwithstanding, for example, *Oldenburgia paradoxa* live long, established in rough cracks. Tiny, low bushes like periwinkle, thyme, and lavender, and members from the family Ericaceae, like cranberries and small types of Erica, are regularly classified as subshrubs (Franklin et al., 2013).

A chamaephyte or midget bush is a plant that produces hibernating buds on continual shoots close to the ground – typically woody florac with perennating buds borne near the ground, generally under 25 centimeters (9.8 in) above the soil surface. The importance of ground closeness is that the buds stay inside the soil surface layer and are subsequently fairly shielded from unfriendly impacts. In like manner, the chamaephyte predisposition is particularly useful in distressing conditions, for instance: biological systems on nutrient lacking soils or rock uncovered highland or lowland environments, where the occasional or lasting breeze and freezing conditions are inclined to kill developing shoots. The term chamaephyte is most officially utilized within the setting of Raunkiær vegetation structures. Instances of chamaephytes are most of the types of the maquis and different plants of submediterranean arid biological systems (species, for example, Rosemary, *Rosmarinus officinalis*, Thyme, and *Thymus vulgaris*); others incorporate African wild olive, heather species (for example *Calluna vulgaris* and Ericas), Chamaephytes like cushion plants, and edelweiss (*Leontopodium alpinum*).

Star Grass

Star grass is a temperate environment toxic weed with a broad structure of crawling rhizomes and stolons. It has a place with the Poaceae family. Rhizomes and stolons are thin intense and layered ordinarily forming roots at the nodes. Most rhizomes are present in the upper 6cm of the soil yet some stretch further. The rhizomes can endure dry season and significant drought, however not extreme sun exposure (Foxcroft et al., 2017).

Stems are woody and less erect, meager and some less spread. The leaf sharp edges are level and adaptable ordinarily around 10cm long. They have long hairs around the shading area, especially on the edge. Inflorescences are umbel-like with typically 4-8 spike-like branches generally 1. 5-3 inches long. The weed in all actuality does well in areas with a temperature range of 18-27°C. It requires full daylight for its ideal development. Precipitation level of 800-1200mm per annum inclines toward its development. It ordinarily seems, by all accounts, to be dry season conducive and subsequently can't endure waterlogging for a delayed number of days. It truly does well in a wide range of soil types going from sandy to weighty mud. Best in lighter soils. They develop in soils with pH of 4.5 to 8.0.

The weed is local to East Africa (Kenya, Ethiopia, Uganda, and Tanzania) in heights of 800-2000M above sea level for the most part along the rift valley. The weed is broad and naturally found in jungles and subtropical. The weed typically proliferates through seeds or vegetative. Rhizomes and stolons are scattered by soil developments. Seeds are typically scattered by water soil development, agrarian tools and other human activities. Seeds ordinarily grow when great conditions prevail. The seeds can remain dormant for as long as 4 years yet most normally sprout after 2 years.

Succulent

Plant with meaty leaves. Succulents prefer direct sun, however assuming yours is laid in a similar spot consistently, all things considered, only one side is getting sufficient light. Succulents will incline near the sun, thus revolving them will assist them in providing an upright stance. Succulent plants, otherwise called succulents, are plants with sections that are thickened, beefy, and engorged, ordinarily to hold water in dry environments or soil conditions (Fahad et al., 2017). It is a trademark that isn't used experimentally for the assistance of most families and genera of plants, since it regularly can be utilized as a precise trademark just at the single-species level. The word succulent comes from the Latin word sucus, indicating 'squeeze', or 'sap'.

Succulent plants might store water in different places, like leaves and stems. The water content of some delicious organs can get up to 90–95%. Others likewise incorporate roots, subsequently, geophytes that endure tough periods by disintegrating back to underground stockpiling parts might be viewed as succulents.

In horticultural plants, the word succulent is occasionally utilized in a manner that excludes crops that botanists consider as succulents, like desert plants. Succulents are regularly developed as decorative plants due to their striking appearance, just as their capacity to flourish with generally negligible care. Many plant families have different succulent species found inside them (around 25 plant families). In certain families, like Cactaceae, Crassulaceae, and Aizoaceae, the majority of the species are succulents (Elmer & White, 2018). The environments of these water-saving plants are typically in regions with high temperatures and low precipitation, like deserts. Succulents can flourish with reduced water sources, for example, fog and dew, which makes them ready to thrive in an environment containing a scant water supply.

Suture

Crease along which units or different fruit or seed products split open. Sutures are commonly found in peas, where they break open once the pod has become dry and release the content into the environment. The various kinds of sutures are based on the type of plant material. Monofilament sutures comprise a solitary string. This permits the stitch to all the more effectively pass through the pod. Multi sutures comprise a few little strings twisted together. This can prompt better protection for the internal seeds.

Symbiosis

Beneficial relationship between at least two organic entities. Any of a few survival methods between individuals from two distinct species, including mutualism, commensalism, and parasitism. Both positive (gainful) and negative (ominous to harmful) associations are subsequently included, and they are called symbionts (De Waele & Elsen, 2007).

Any relationship between two species that live respectively is harmonious, regardless of whether the species advantage, hurt, or have no impact on each other. Advantageous plants, or the course of beneficial interaction, is when two plants live amicably in some sort. There are four sorts of beneficial interactions - mutualism, commensalism, endosymbiosis/ectosymbiosis, and parasitism. The term symbiosis arises comes from the Greek terms 'with' and 'living'. They define a connection between two species categories or organic entities that can be useful for the two players. Commensalism is the place where one organism benefits from the relationship, however, the other is not benefiting, or hurt by it. An illustration of commensalism would be a spider forming a web in a tree. It doesn't hurt the tree, however, it helps the bug. Parasitism occurs when one organic entity benefits to the detriment of the other. An illustration of parasitism is an aphid benefiting from the sap of a plant. The aphid benefits, however, the plant is worse off. In the plant world, the best harmonious relationship is mutualism - where the relationship benefits both.

Shrub

A woody extending plant with stems emerging at or close to the ground. A shrub is a little to medium-sized all-year woody plant. Different from herbaceous plants, they have woody stems over the ground. Bushes can be deciduous or evergreen. They are unique from trees by their various stems and more diminutive stature, under 6–10 m (20–33ft.) tall. Small shrubs, under 2 m (6.6 ft.) tall are named subshrubs. Numerous plant groups have species that are shrubs, and others that are trees and herbaceous plants. A few definitions define that a bush is under 6 m (20 ft) and a tree is 6 m (Chen & Yamaguchi, 2002). Others give 10 m (33 ft.) as the cut-off point. Many types of tree may not arrive at this full-grown stature due to poor growth conditions, and look like a shrub-based plant. Anyway, such species can possibly become taller under the best developing conditions for that plant. As far as life span is concerned, most bushes fit in the middle of perennials and trees; some may just go on around five years even in good conditions, others, normally the bigger and woodier, may live to at least 70.

Shrubland is a natural land defined by shrubs; there are numerous unique forms all over the planet, such as shrub steppe, shrub bog, moorland, fynbos, and maquis. In green parks, a region to a great extent is filled with shrubs known as a growth, shrub boundary or shrub nursery. There are many nursery cultivars of shrubs, reproduced for blossoming, for instance, rhododendrons, and in some cases leaf tone or shape. Contrasted with trees and herbaceous plants, maybe a moderately modest number of shrubs have agrarian or corporate employments (Clemens et al., 2013). Aside from a few berry-bearing plant categories, few are eaten directly, and they are for the most part excessively minor for lumber use. Those that are utilized incorporate a few perfumed varieties like lavender and rose, and a wide range of plants with restorative abilities. Tea and coffee remain on the tree-shrub border; they are ordinarily collected from shrub-sized plants, yet these would turn out to be small trees whenever left to develop.

Stalk

The slender stem-like product that connects the blossom, leaf or organic fruit to the plant or tree. The petiole comprises a stalk that connects the leaf to the plant stem. For petiolate leaves, the leaf stalk might be extensive, like the leaves of rhubarb and celery, brief or totally missing, in which case the edge connects directly to the stem and is supposed to be sessile. Subpetiolate leaves contain short petiole, and may seem sessile. The broomrape group Orobanchaceae is an illustration of a family whereby the leaves are consistently sessile. In other plant groups, like the speedwell variety Veronica, petiolate and sessile leaves might happen in various species. In the grasses (Poaceae), the leaves are apetiolate, however, the leaf edge might be restricted at the intersection with the leaf sheath to shape a pseudopetiole, as in Pseudosasa japonica. In plants with compound leaves, the stalks are joined to a continuation of the petiole called the rachis (Craigie, 2011).

Every leaflet might be connected to the rachis by a brief stalk called the petiolule. There might be enlarged locations at one or the flip side of the petiole known as pulvina, made out of an adaptable tissue that permits leaf development. Pulvina are common in the bean family Fabaceae and the plant family Marantaceae. A pulvinus on a petiolule is known as a pulvinulus. In certain plants, the petioles are straightened and broadened to become phyllodes (otherwise known as phyllodia or cladophylls) and the actual leaves might be diminished or missing.

Subsequently, the phyllode functions as the elements of the leaf. Phyllodes are regular in the Acacia genre, particularly the Australian type, occasionally found in the Acacia subgenus Phyllodineae. Acacia koa containing phyllode along the branch and the composite leaves in Acacia koa, the phyllodes are tough and thick, permitting the tree to endure stress conditions (Chandler & Sanchez, 2012). The petiole permits limited hydrophytes to have leaves drifting at various depths, the petiole being between the node and the stem. In plants like rhubarb, celery, artichokes, and cardoons, the petioles "stalks" are developed as edible plants. The rhubarb petiole develops directly from the rhizome and creates the leaf at its edge. Botanically it is classified as a vegetable or culinary consumed as a natural fruit.

Sagittate

Sagittaria longiloba is a type of blossoming plant in the water plantain group known by the name longbarb arrowhead. It is local to the southwestern and central United States in states like Texas, Kansas, Nebraska, New Mexico, Oklahoma and Arizona in addition to Mexico, Venezuela and Nicaragua. It grows in sluggish, stale, and vaporous water bodies like lakes and little streams, and occasionally developed territory, for example, rice fields and water system ditches. Sagittate is a perennial water plant developing from a round tuber. The leaves are sagittate, or formed like sharpened stones with two longer, smaller, pointed flaps opposite the more limited tip (Bourgaud et al., 2001). The leaf edges are borne on extremely long petioles. The plant is monoecious, with individuals bearing both male and female blossoms. The inflorescence which transcends the outer layer of the water is a raceme comprised of a few whorls of blossoms, the most limited node bearing female blossoms and upper nodes bearing male blossoms. The blossom is up to 3 centimeters wide with three white petals. The male blossoms have rings of stamens at the focal point. Female blossoms each have a circular bunch of pistils that form into the head of tiny fruits.

Salicoid

A leaf tooth where the average vein closes in dull but not murky conditions, no laterals are involved, maybe near a theoid tooth, occasionally working as an extrafloral nectary, such as cucurbitoid, cunonioid, begonioid, chloranthoid, dillenioid, malvoid, spinose, urticoid, violoid, monimioid, platanoid, and rosoid.

The Salicaceae is the willow group of blooming plants. The family Salicaceae incorporates the willows, poplar, aspen, and cottonwoods. Considered to be firmly connected with the Violaceae and Passifloraceae. Under the new circumscription, all individuals from the family are trees or bushes that have standard leaves and are typically deciduous (Arruda et al., 2015). Most individuals have serrate or dentate leaf edges, and those that have such toothed edges all display salicoid teeth; a salicoid tooth being one in which a vein enters the tooth, grows, and ends at or close to the peak, close to which are circular and glandular bulges called setae. Individuals from the family frequently have blossoms that are diminutive and subtle, and all have ovaries that are unique or half-substandard with parietal placentation.

Salicylic Acid

Orthohydroxybenzoic acid, derived from the bark of the wintergreen leave and white willow, its salts are called salicylates. Salicylic acid is a beta-hydroxy corrosive. It's notable for lessening skin acne by peeling the skin and keeping pores clear. You can find salicylic acid in an assortment of over-the-counter (OTC) plant-based cosmetics. Salicylic acids are best for gentle skin inflammation (clogged pores and whiteheads). It can likewise assist with forestalling future breakouts. When your hair follicles (pores) get jammed with dead skin cells and oil, zits, whiteheads, or pimples (pustules) regularly show up (Zehnder et al., 2007).

Salicylic acid enters into your skin and attempts to break up the dead skin cells blocking up the pores. It can require half a month of utilization to see its full impact. Your primary care physician or dermatologist will suggest a structure and measurements explicitly for your skin type and your skin's present condition. They may likewise suggest that for 2 or 3 days, you just apply a restricted quantity to a little space of impacted skin to test your response prior to applying to the whole region.

Salverform

An example of a polysymmetric corolla, salver-or trumpet-shaped, similar to the corolla of Primula with a long, thin cylinder and highly extended level, spreading appendage. Salverform blossoms are formed like a long cylinder, and have a smoothed circle of petals at the top. There's hardly any bending of the petals, so the substance of a salverform blossom is level. Salverform blossoms are antinomorphic, possessing spiral balance.

Samara

A dry, indehiscent fruit product with its wall partitioned into a wing or wings, a variation of an achene or a mericarp of a schizocarp. A samara is a winged achene, a kind of fruit product wherein a smoothed wing of sinewy, papery tissue forms from the ovarian wall. A samara is a dry plant product, and is indehiscent (not opening along a crease). The state of a samara allows the breeze to divert the seed farther from the tree than normal seeds would go, and is subsequently a type of anemochory (Yamada et al., 2003).

Occasionally, the seed is in the focal point of the wing, as in the elms, the hoptree, and the bushwillows. In different cases the seed is on one side, with the wing reaching out to the opposite side, making the seed autorotate as it falls, as in the maple trees.

Sanio's Law

Karl Gustav Sanio (5th December 1832 to 3rd February 1891) was a Prussian plant scientist and an educator of herbal science at the Königsberg University. He noticed patterns in the development of plant vasculature and wrote a few articles on wood and cambium. He analyzed examples in the size transmission of the xylem vessels in five explanations known as Sanio's laws, it was additionally among the first to portray the structure of pressure wood by conifers (Yakhin et al., 2017). In 1872 he distributed a number of perceptions on an example in the size of xylem tracheids, expanding in size from the highest point of a tree to a specific height, later which they decline again to the foundation of the tree. He communicated these as five examples that go by the name of Sanio's laws. Sanio's Laws are mostly considered to constitute coniferous woods:

- Tracheids grow in size from within to outside the wood (as found in annual rings) in stems and branches. Along these lines, they increase with age to a pinnacle and afterward remain consistent.
- The constant size shifts in the stem, growing upwards but lessening later a specific stature.
- The steady size in branches is more modest than in stems connecting to them
- In contorted branches, the size of tracheids in external rings can be more modest.
- In the development rings of the roots the width of the tracheid increases and diminishes prior to rising again to a steady size

Sanio was additionally quick to take note of the arrangement of wood of high thickness under unique circumstances. The bryophyte variety Sanionia was named after him. The scientist name abbreviation Sanio is utilized to show this individual as the creator while referring to a plant name.

Saponins

Water-solvent and occasionally harmful glycosides dependent on steroid and terpenoid alcohols, become a froth when in fluid and afterward shaken. Saponins include an enormous group of primarily related mixtures comprising a steroid or triterpenoid aglycone (sapogenin) connected to at least one oligosaccharide moieties (Yakhin et al., 2017). They are described by their hemolytic action and frothing properties and are liable for the harsh taste and astringency to plant materials containing a high centralization of saponins. Saponins are weak absorbents. The vast majority of the saponins develop insoluble edifices with 3-b-hydroxysteroids and are known to interface and develop blended micelles with bile acids and cholesterol.

Despite the fact that saponins bring down cholesterol in some living species, the hypocholesterolemic impacts of saponins are more speculative. Moreover, they develop insoluble saponin–mineral complexes with iron, zinc, and calcium. The most well-known saponins in vegetables include the soyasaponins, which are characterized into A, B, and E saponins based on the compound of the aglycone (Wisler & Norris, 2005). They are known to be an intense calcium-enacted potassium channel opener, and as such can be utilized for treating cardiovascular, urological, respiratory, neurological, and different problems. Saponins have been accounted for in numerous edible vegetables. The significant sources of dietary saponins are vegetables, and many forms of saponins can be available in a similar bean. Saponin content might change even among similar types of palatable vegetables.

Sapwood

Figure 167: Sapwood is the hard external part of wood.

Source: https://infocabin.net/difference-heartwood-sapwood/

The external and innate practical part of the wood of a trunk or enormous root, c.f. heartwood. Sapwood, likewise called alburnum, external, living parts of the supplementary wood of trees, which take part in transportation of water and minerals to the tree crown. The cells thusly contain more water and do not have the deposits of obscure chemical substances generally found in heartwood. Sapwood is accordingly paler and gentler than heartwood and can ordinarily be defined in cross segments, as in tree stumps, albeit the extents and peculiarity of the two kinds are variable in unique species (Weitemier et al., 2014).

Sarcotesta

The sarcotesta is a beefy seedcoat, a form of testa. Instances of seeds with a sarcotesta are pomegranate and some cycad seeds. The sarcotesta of pomegranate seeds comprise of epidermal cells derived from the integument, and there are no arils on these seeds. A beefy testa, or a plump layer of the testa (mesotesta), external to the sclerotesta, when the last option is available, c.f. additionally aril, arillode, caruncle, elaiosome.

Stoloniferous

In plant science, stolons are horizontal relations between living beings. They might be necessary for the living being, or of its skeleton; ordinarily, stolons are outside the living organisms. In plant science, stolons are stems that develop at the soil surface or just subterranean that form unusual roots at the nodes, and new plants from the buds. Stolons are regularly called sprinters (Wilkinson et al., 2012). Rhizomes, conversely, are root-like stems that may either develop on a level plane at the soil surface or in different directions underground. Thus, not all flat stems are called stolons. Plants with stolons are called stoloniferous. A stolon is a plant proliferation method and the complex structures formed by a mother plant, including every one of its clones created from stolons forming a solitary hereditary individual, a genet.

Stolons from the corm of a Crocosmia are stems that arose out of axillary buds at the nodes of the tunic leaves. Stolons could possibly have long internodes. The leaves along the stolon are generally tiny, however in a couple of cases, for example, *Stachys sylvatica* are ordinary in size. Stolons emerge from the foundation of the plant. In strawberries the base is over the soil surface; in numerous bulb-style species and plants with rhizomes, the stolons remain underground and develop shoots that ascent to the surface at the closures or from the nodes. The nodes of the stolons produce roots, regularly all around the node and chemicals delivered by the roots make the stolon start to shoot with ordinary leaves. Typically, the stolons pass on after a short while, while rhizomes endure ordinarily for a longer time or for the duration of the plant, adding more length every year to the terminals with dynamic development (Waraich et al., 2011). The uniform development of stolons results from the exchange of various chemicals delivered at the developing point and chemicals from the primary plant, for certain situations the stolon and rhizome development are impacted by how much light the plant gets with increased production and growth from plants with partial shade and sun, while plants in full daylight sun or all shade produce less stolon.

Snow bell

The Japanese Snowbell is a deciduous blossoming tree with a somewhat contracted propensity that can be jar molded when young, however with low level fanning and an all the more extensively adjusted crown with age. It inclines toward soggy, acidic well-soaked soil in the full sun to partial shade. It projects a light shade and the tree is tolerably salt lenient.

The plant's hard wood has been utilized to form the ribbing for umbrellas just as toys. Its branches were utilized for strolling sticks. It blossoms in pre-summer and its fruit product ages in pre-fall on the successive season's development. The foliage happens on upper piece of branches and doesn't block pendulous blossoms altogether. The bark has random, orange-earthy coloured gaps which adds interest to a colder time of year. This is an amazing adaptation for a slope or inclined environment. Humidity is essential in growing them, don't permit the soil to dry out (Van Dijk et al., 2021).

Plants close to a deck, in a yard, a boundary or in tiny groups. Plant it close by a structure on the northern part to shield it from strong breezes and to give a delightful dappled shade to under plantings. It doesn't endure dry environments very well overall. Winter pruning should be possible to control the state of this tree. Regularly, this variety will self-seed. Seedlings may be found under more mature trees.

T

Taproot

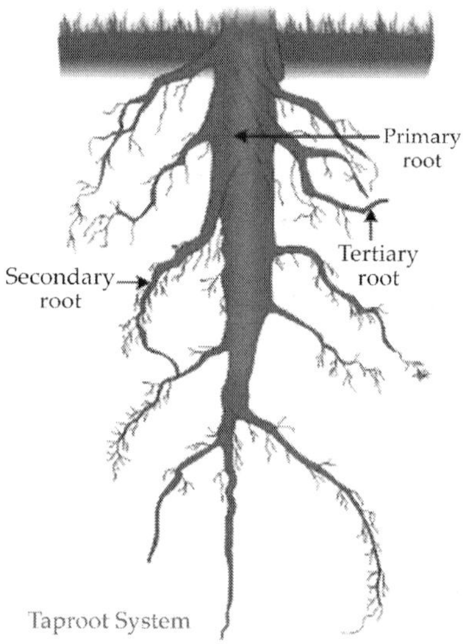

Figure 168: Taproots are mostly found in deciduous plants.

Source: https://vivadifferences.com/understanding-taproot-vs-fibrous-root-14-differences-plus-examples/

Taproots are primary descending roots of a plant having a single dominant roots axis. Taproots normally sprout laterally and tend to be straight. Taproots are thick and grow directly downwards. Taproots play various functions in plants. For example, in carrots, the taproot works as a storage organ. Plants with taproots have a taproot system that may start off with a main root and develop some branching structures afterward. There are different types of roots. There is the conical root where the taproot has a conical shape. For the napiform root, the taproot has a top-like appearance while in the fusiform root, the taproot is characterized by a wide middle area and tapers towards the top and bottom (Van Bruggen et al., 2016).

Tartareous

Tartareous is a term used in botany to refer to plants with a surface that is crumbling, rough, thick and course. This is very advantageous to plants as it gives them some sense of protection from mechanical damage among other things. In some cases, this could be an indication that the plant is sickly or very healthy. It all depends on the nature of the plant.

Taxon

Taxon is a term used in botany to refer to category or group in a system of biological classification. At taxon is also a group or population of plants that are usually inferred to be phylogenetically related and have some common characteristics. A taxon is usually identified by a particular name. The ranking is also useful in identifying a taxon (Van Breusegem et al., 2001). A ranking is obtained once a group has been accepted and established. Nomenclature codes are used when a taxon is given a scientific name. When the codes are used, it specifies the particular grouping for certain plants. The eight main ranks include the species, genus, family, order, class, phylum, kingdom, domain and life.

Taxonomy

Taxonomy is the study of practices and principles of classification. It studies how organisms are classified into groups based on their shared characteristics. Here plants are grouped into taxa. The taxa are then given a taxonomic rank. Under taxonomy, the principle ranks are the domain, kingdom, phylum, class, order, family, genus and species. Over the years, taxonomy systems have been improved leading to the development of a ranked system that is the Linnaean taxonomy used in categorizing organisms and binominal nomenclature used in naming organisms. Taxonomy is used by taxonomists in making taxonomic classifications.

Tegmen

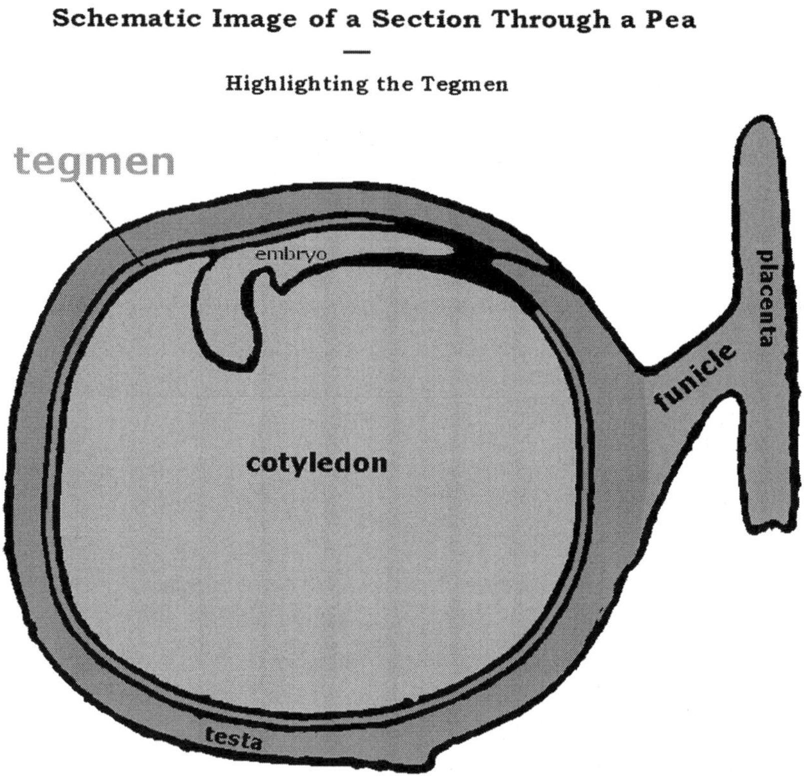

Figure 169: The Tegmen is the inner layer of the seed coat.

Source: https://en.wiktionary.org/wiki/tegmen

Tegmen is the inner layer of the seed coat. The seed coat is also known as the testa. Tegmen is formed when the inner integument of the ovule develops. Normally, the Tegmen is thin in nature, membranous and whitish or hyaline in color. The Tegmen is very important for seeds as they help protect the embryo from mechanical damage and dehydration (Van der Kooi & Ollerton, 2020). Tegmen have two layers namely the exotegmen and the endotegemen. There are some seeds that do not have Tegmen and are therefore called unitegmic seeds. There are similarities and differences between the Tegmen and testa.

Tendril

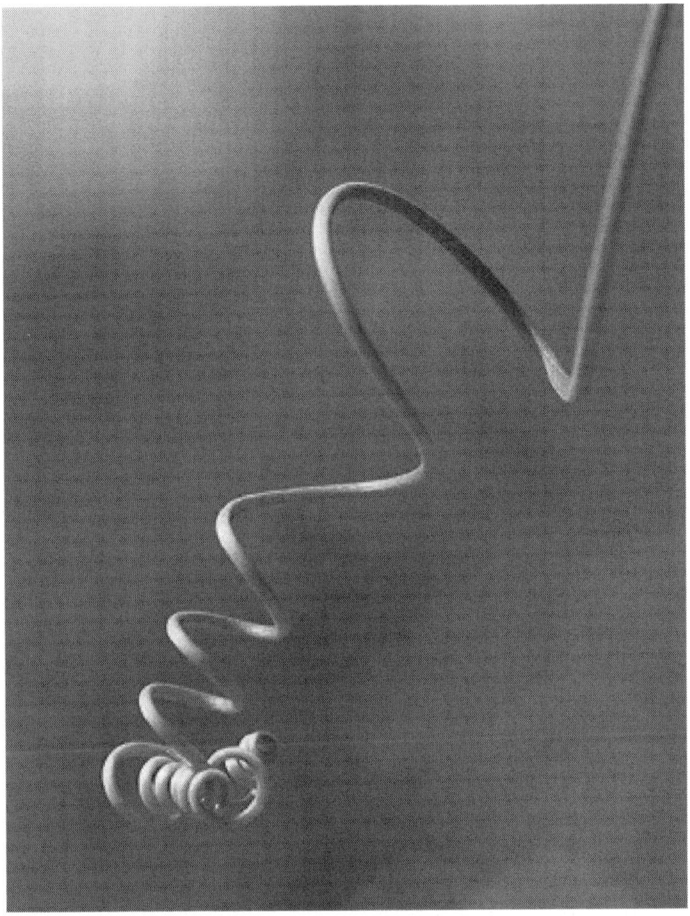

Figure 170: Tendrils allow climbing plants to attach themselves to objects.

Source: https://www.britannica.com/science/tendril

Tendrils are slender organs that have been modified from a stipule, leaflet, leaf or stem. Tendrils are threadlike appendages of climbing plants that enable them to cling on to objects. Tendrils are very important for climbing plants they give them support and attachment. In most cases, tendrils are responsive to chemical factors and touch by twining and curing among other things. For plants like the garden pea, leaflets are modified to become leaflets while in the yellow vetch the entire leaf is modified to become tendrils.

Tepal

Figure 171: A tepal is a segment of a perianth.

Source: https://www.youtube.com/watch?v=xjsLVurtOTk

Tepal is a term used in botany to refer to a segment of a perianth which is either a petal or a sepal. The tepal is used when the perianth segments cannot be distinguished based on appearance. The tepal is often used when the perianth is undifferentiated. Tepals can be described using the following terms puberulous, puberulent, downy, soft hairs, short or dense fine. Terms used when describing leaves are also used when describing the shape of a tepal. It is believed that the undifferentiated tepals are the ancestral conditions in flowering plants (Twidwell et al., 2013).

Terete

Terete is a term used in botany when describing the cross-section that is like a distorted circle with a single surface surrounding it. Terete is usually contrasted with cross-sections that are flattened and the upper surface is distinct from the lower surface. If the section of a branch is round or almost round the branch is said to be terete. For succulents with fleshy leaves, they are said to be terete. Fruticose lichens are examples of terete plants as they have a roughly circular cross-section and cortex. For plants that are not terete, they are referred to as subterete.

Terminal

Terminal is a term used in botany when referring to buds located at the end of an embryonic shoot. Terminal buds are mostly located at the tips of shoots or the apex. For plants with terminal buds, the resources are concentrated at the bud. This is done to foster growth in the plant and also to enable it to grow taller. The terminal bud may be located at the end of the main stem. The terminal bud tends to appear at the end of the main shoot of plants.

Ternate

Ternate is a term used in botany to refer to things arranged in groups of three. It could be used to refer to leaves arranged in whorls of three or a single leaf with leaflets in groups of three. An example of a plant with ternate leaves is poison ivy. A single leaf of poison ivy is divided into three portions that are the leaflets (Taiz et al., 2015). As the leaflets are three, they could also be referred to as trifoliate. Other examples of plants and trees that are ternate include sugar maple, Amur maple, white clover and shamrock. For ternate leaves with whorl arrangement, the leaves are connected at the node.

Terrestrial

Terrestrial refers to a habitat located on land, rocks or other plants. Lithophytic is a habitat on rocks while epiphytic is a habitat on plants. The term can also be used to refer to an ecosystem hence the name terrestrial ecosystem and it refers to an ecosystem found on landforms. Terrestrial plant is a term used in botany to refer to plants that grow on land rather than on rocks, trees or water. Terrestrial plants make use of resources available on land to germinate and reproduce.

Figure 172: Terrestrial plants live on earth.

Source: https://lisbdnet.com/what-are-terrestrial-plants/

Testa.

Is also known as the seed coat. It is the outermost skin layer, usually hard on the seed. Testa may be light such as in corn or dark such as in beans. It has a number of functions. First and foremost, it protects the embryo of the plant as well as the endosperm from damage (Szpak, 2014). Secondly, it plays a key role in aiding the seed to mature into a plant by protecting it from freezing, heat, water that can potentially affect the maturity stages of a seed. It also insulates the seed from sub-freezing temperatures. Thirdly, it prevents seed damage due to attack by microbes in the soil or when being inserted into the ground.

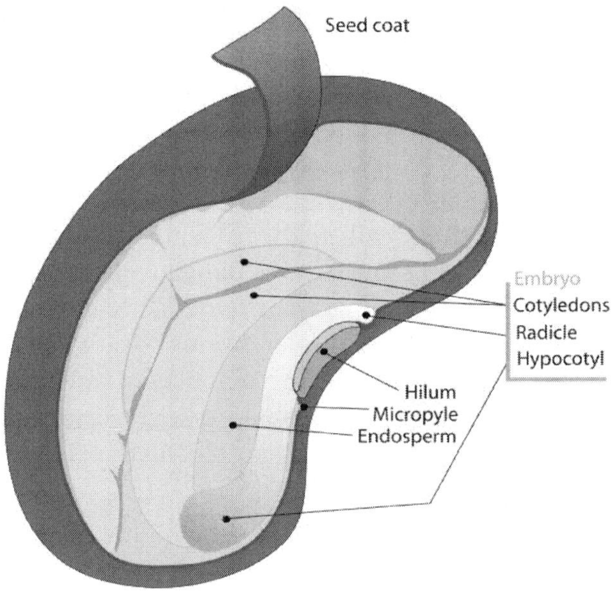

Figure 173: The testa is the outer covering of seeds.

Source: https://pediaa.com/what-is-the-difference-between-testa-and-tegmen/

Tetrad

Tetrad refers to a group of four. The term tetrad is mostly used when referring to four pollen grains that remain fused through maturity. The pollen grains are produced through meiosis where four cells are produced by meiotic cell division. For some plants, the cells are haploid. The four haploid cells are formed in a tetrahedral fashion are a result of meiotic microsporogenesis.

Based on pollen grains arrangement, there are five types of tetrads that have been recognized. They include linear tetrads, tetrahedral tetrads, rhomboidal tetrads, tetragonal tetrads and decussate tetrads (Stenberg, 2017).

Tetragonal

Tetragonal is used in botany to imply having four corners. It is mostly used when referring to the cross-sections of stems of herbaceous. It can also be used to refer to a four-sided anther. It can also be used to refer to plant cells. The symmetry of the plant does not affect the functioning of the plant. For cells, by being tetragonal makes it to be the least consuming. It is often used by researchers when simplifying other types of crystalline patterns. Lamiaceae is an example of a plant whose stem cross-sections are tetragonal in nature.

Tetramerous

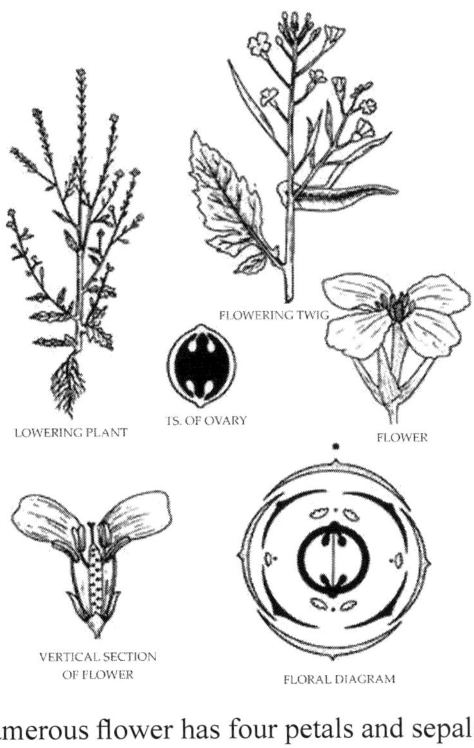

Figure 174: A tetramerous flower has four petals and sepals.

Source: https://www.quora.com/What-is-a-tetramerous-and-pentamerous-flower

Tetramerous is used in botany refers to flowers with four parts hence the name tetramerous flower. For the tetramerous flower, the floral parts are arranged in four or multiples of four. In this case, the whorl of flower parts have four flower parts. A good example of such a flower is *Brassica juncea*. The tetramerous flower has four stamens, four petals, four sepals and two carpels. In some cases, one of the whorls may have a different number of floral parts.

Tetraploid

Tetraploid refers to a cell having four complete sets of chromosomes in every cell. It is a phenomenon under ploidy which refers to the number of complete sets of chromosomes in a given cell. The sets of chromosomes are related to the number of paternal and maternal chromosomes in a chromosome pair. In plants, a large number of polyploids are tetraploids. For tetraploid plants 2n= 14. This is a result of the fusion of diploid gametes. There are plants that are tetraploid due to meiosis while others are through mitosis. Wheat is an example of a tetraploid plant (Singh, 2015).

Tetraspore

The Tetraspore is the asexual spore of red algae. The asexual spore is thus given the name Tetraspore because the sporangium produces four spores. The tetrasporophytic phase in the life algae as a result of meiosis brought about the Tetraspore. The tetrasporangium produces the tetraspores and also hosts them as they germinate. The germinated tetraspores produce a sexual haploid generation. A carpospores forms the ground for the development of tetrasporophyte. The tetraspores become either male or female gametes after germination. In algae, the male gametophyte produces spermatangial branches that contain the male gametes while the female gametophytes produce carpogonial branches.

Thalamus

Thalamus also called the receptacle is a part of a flower found near the stalk of the flower. Usually, it is the part of the stem that is thickened.

Figure 175: The receptacle is also called the thalamus.

Source: https://www.quora.com/How-do-you-define-the-thalamus-of-a-plant

It is from the thalamus that other structures of the flower grow. The thalamus is therefore a modified stem exhibiting distinct nodes and internodes. The thalamus also plays the vital role in supplying flowers with nutrients. Thalamus is also used in botany to refer to an inflorescence disk in members of the Asteraceae or a calyx in the Carl Linnaeus.

Thallus

The thallus is the vegetative part of a lichen that has both photobiont and mycobiont. Its plural is thalli. Plants having thallus were initially called thallophytes. Organisms or structures that resemble a thallus are called thallose, thalline, thalliform, thallodal or thalloid. Organisms having the name thallus are usually non-moving and have no organization in their tissues. Even though there is no organization in their structures, they have distinct plants such as the stems, roots and leaves. Thallus is largely undifferentiated but can be identified as there are functional and visible differences. The kelp is an example of a thallus (Schuurink et al., 2006).

Theca

Theca is according to botany refers to the anatomy of a plant. For instance, in angiosperms, the theca is consists of a pair of microsporangia. The microsporangia are usually adjacent to each other and have a stomium. The stomium is the common area of dehiscence of the microsporangia. The anthers bear the microsporangia. The bilocular is a typical anther that consists of two thecae and each of the theca consist of two microsporangia. The two microsporangia are known as pollen sacs. For a male strawberry plant, the anthers have parallel thecae. Thecae is the plural of theca.

Thorn

A thorn is a sharp, stiff point modification of a stem. Thorns cannot be detached from the plant without tearing the subtending tissue. In most plants, thorns are a modification of the axillary shoot stem for which leaves are reduced and have determinate growth. Thorns help protect the plant from herbivores. Examples of plants with thorns include Citrus, anchor plant, honey locust and bougainvillea. Cactus is a plant whose leaves have been modified into thorns. This is very useful in reducing the amount of water lost due to evaporation. For plants in the Apocynaceae, the axillary buds develop as thorns.

Figure 176: Thorns help protect plants from herbivores.

Source: https://www.istockphoto.com/photos/thorn

Throat

Throat is the opening of a perianth or corolla. The throat allows pollen grains to be released from the anthers. The throat may have hairs. If there are several hairs, there is a higher chance that pollen grains will remain in the flower. Several hairs also reduce the ability of pollinators to get into contact with pollen grains. For throats that have reduced numbers of hairs, a large number of pollen is released into the atmosphere (Smith & Knapp, 2001).

Thyrse

Thyrse is a type of inflorescence characterized by the main axis growing indeterminately while the branches have determinate growth. The thyrse is also called thyrsus. This kind of inflorescence is mostly noted in grapes. Thyrse is an example of mixed inflorescence with a paniculate appearance. They may be pyramidal or oblong in shape. Thyrse may consist of a panicle of cymes. For an inflorescence to be termed as thyrse if they have met two conditions. One, the cymes or their branches are indeterminately arranged on the main axis. Secondly, if it is composed of many determinate triflowered cymes. In banana plant, its thyrse has side branches that develop to become the hands of bananas. For *Vitex angus-castus* its inflorescence also has branches.

Tomentum

Tomentum is a dense covering of short and matted hairs on plant leaves. They are usually called plant trichomes. Other names of tomentum are hairy and tomentous. Tomentum is different in different plants but they cannot be separated. Tomentum sever a number of functions in plants. They provide both chemical and physical protection for the leaf against microbial organisms and insects. It also prevents excessive water loss in leaves through transpiration by providing a maintenance layer of still air on the leaf surface (Simon & Petrášek, 2011).

Toothed

Toothed is a term used to define the margins of leaves that may be regularly incised. The American chestnut, American Elm, American Beech, Bigtooth Aspen, Eastern Cottonwood, European Linden, European White Birch, Gray Birch, River Birch, Paper Birch, Quaking Aspen and weeping willow are examples of plants with toothed leaves. Research shows that plants have toothed leaves to enable them to cope with cold conditions. This is very important as the tips of the toothed leaves are packed with tissues that help in the transportation of nutrients and water. When water is lost at the tips, it causes water to be pulled out from the roots.

Torus

Torus is the part where a stem fuses fuses with the flower. It is the dense section of a stem from where a flower arises. Accessory fruits, like strawberries and apples, develop from the torus.

Transmitting tissue

Transmitting tissues are specialized tissues that are present in the style of the carpel. Transmission tissues play a vital role during the fertilization of female and male gametes. It makes a pollen tube that grows down the style and connects the inside of the ovary to the stigma. This allows pollen grains to find their way to the ovary allowing fertilization to take place (Rusman et al., 2019). Transmitting tissue normally lies beneath the stigmatic tissue as a band. It also forms the central strand of the style. The transmitting tissues have certain glycoproteins that are useful in stimulating the growth of the pollen tube.

Tree

Figure 177: There are different kinds of trees all specially adapted to survive to a given condition.

Source: https://unsplash.com/s/photos/trees

Trees are perennial plants with elongated stems or trunks that support leaves and branches. The definition of a tree may be narrowed down on the type of growth, height and uses. Different species of trees have evolved over the years to enable them to survive in different environments. In most cases, trees have grown taller to enable them to get sufficient amounts of sunlight. Most species of trees are long-lived with some dating to about a thousand years old. The trunks of trees contain woody tissue which gives them their strength. They also have vascular tissue which allows the transportation of both water and minerals from one part of a tree to another. The trunk of trees is surrounded by bark which is its protective layer (Reid et al., 2009). The roots of trees branch and spread out widely. The roots play a vital role in anchoring the tree and allow the extraction of water and minerals from the soil. The branches of trees divide into smaller branches and shoots. The shoots bear the leaves which are actively involved in photosynthesis as they have chlorophyll which enables them to capture sunlight. Photosynthesis allows the provision of food that will allow the growth and development of the tree.

Trees reproduce by using seeds. Some trees may produce flowers while others such as conifers have seed cones and pollen cones. Tree ferns produce spores. Trees have continuous growth and have different periods of dormancy. Trees are very different. For instance, deciduous trees may not have leaves for the most part of their lives.

Triad

Triad is a group of three. There are various parts of plants that occur in three. The flowers of plants can occur in three in that they have three petals. The leaves of plants can also occur in three in that three leaves can be joined together in a node.

Triangular

Triangular refers to planar having three sides. A number of trees are known for their triangular leaves. *Oxalis triangularis* , also called false shamrock is a plant known for its triangular leaves and molted green color. It is grown as a house plant and is photophilic. Their triangular leaves fold up during the night. The mile-a-minutevine is a plant known for its distinct triangular leaves. Its leaves are equilateral and do not have lobes like *Oxalis triangularis* (Poff & Zimmerman, 2010). Other than plants, there are species of trees known for their triangular leaves. They include black poplar and

Silver Birch. The leaves of plants may be triangular but with different kinds of margins and arrangements in their structure. By having triangular leaves, the leaf surface is broad increasing the area for photosynthesis.

Tribe

Tribe is a taxonomic group ranked between genus and family. In the taxonomic rank, a tribe is above the genus and below the family and subfamily. Like all other ranks above genus, the tribe and sub-tribe is capitalized. When writing names while including the tribe, the standard ending of the name ends with "-eae". Examples of tribes are the Hyacintheae and the Acalypheae. Most tribes have their own subtribes. For instance, for the tribe Hyacintheae its subtribes include the subtribe Massoniinae. The botanical ending of subtribes is the "-inae" as illustrated by Massoniinae.

Trichome

Trichome is any hair-like outgrowths from the epidermis in non-filamentous plants. Trichomes are also appendages on lichens, algae and plants. Trichome is of diverse structure and function. Hair coverings on plants are referred to as indumentum while surfaces bearing them are said to be pubescent. There are two main groups of trichomes on plants namely algal trichomes and plant trichomes. The algal trichomes are found on algae. They are elongate hair-like structures produced by terminal cells. Trichomes in algae play a vital role of forming a sticky network that helps maintain soil structure (Pauchard et al., 2009).

In plants, trichomes have different features that vary between the organs of individual plants and the species of plants. The features affect subcategories in which trichomes are grouped. Some of the defining features of trichomes include the presence of cytoplasm, multicellular or unicellular, straight or hooked and eglandular or glandular. Trichomes play a vital role in protecting the plant from a large range of detriments such as transpiration and UV light. Bean leaves have trichomes which are useful in helping the leaves capture insects by impaling their feet. For the European wool carder bee, trichomes on plants are used in building their nests. They scrape trichomes of plants and use them to line their nest cavities.

Trifid

Trifid means split into three parts. A plant can be considered a Trifid as it can be subdivided into three main parts made up of the stem, leaves and roots.

Each part plays its own role in ensuring the growth, development and health of the plant.

Trifoliate

Trifoliate is a compound leaf made up of three leaflets. Trifoliate leaves are also called ternate or trifoliolate. There are different species of plants with trifoliate leaves. The trifoliate leaflets are attached to a common stalk. Plants under the family Rutaceae are known to have trifoliate leaves. Examples of plants with trifoliate leaves include *Acer griseum*, *Acer triflorum*, *Butea* spp., *Eleiotis* spp., *Medicago truncatula* and *Nandina domestica* among others.

Trifoliolate

Trifoliolate is a compound leaf with three leaflets attached to a common stalk. Trifoliolate leaves are also called trifoliate or ternate leaves.

Trimerous

Trimerous is a term used in botany to refer to flowers with three parts in a given whorl. Trimerous plants have three petals in each whorl. Monocots plants (plants with a single cotyledon) are known to have Trimerous flowers (Prajapati & Modi, 2012). Plants with Trimerous flowers are known to have fibrous root systems and parallel leaf venation. Trimerous flowers have three sepals, three petals, three stamens and three carpels in their flowers. Examples of plants with Trimerous flowers include amaryllis, tulips, bluebells, orchids, iris, daffodils, lilies, ginger, banana, palm, bamboo, maize, rice, grass, corn, wheat, onions and garlic.

Trinerved

Trinerved is a term used in botany to refer to leaves with three veins or nerves. The nerves play a crucial role in ensuring both water and mineral salts to the leaves. Among the plants that are Trinerved is *Alsinidendron trinerve* that is a small branching shrub under the pink family. The shrub has thick fleshy trinerved leaves that are long. The flowers of the plant are borne in open dusters and do not have petals.

Triplinerved

Triplinerved is a term that refers to leaves with three main nerves with lateral nerves arising from the mid-nerve above the base of the leaf. *Ayapana triplinervis* is an example of a plant that is triplinerved. Other than its leaves being 3-nerved, they are glabrous, subentire, acuminate, lanceolate and purple in color. The leaves of the plant contain volatile and Ayapana oil. The leaves are normally used as a stimulating medicine (Neumann et al., 2009).

Triporate

Triporate is a term used in botany to refer to pollen. It is among the types of colpate pollen grains in angiosperms. The Triporate pollen grain has three furrows on their surface and is commonly found in dicots where as its opposite which is the monoporate are found in monocots. *Oenothera speciosa* is an example of a plant with Triporate pollen grains. Eudicots are three colpied as their shapes have been evolutionarily been derived from the tricolpate pollen. The same evolutionary trend is also noted in other orientations of furrows that classify pollen as either colpate or sulcate.

Triquetrous

Triquetrous means more or less triangular in cross-section but acutely angled plants or fruits. It means that the part has three distinct longitudinal ridges. There is the three-cornered garlic whose botanical name is Allium triquetrum. It is a bulbous flowering plant in the genus Allium. The leaves of the plant have are quite fleshy and have three sides hence its botanical name. The leaf has three sides, smooth and have different sizes. The leaf also has a cluster at the base.

Trivalve

Trivalve refers to plants with thee valves. It is also called trivalvar. An example of such a plant is the Zootrophion Trivalve.

Trivial name

The trivial name, also known as the common name is the second word in the two-part scientific name of an organism. The trivial name is usually based on the normal language of everyday life. The trivial name is often compared to the Latinized part of the scientific name. The trivial name was created to enable members of the public to be able to refer to a particular species of plants without having to pronounce or memorize the Latinized specific name. For most species of plants, the trivial name tends to be superficially similar to the scientific name of the organism (Nikunj et al., 2014).

Trophophyll

Trophophyll refers to the vegetative nutrient producing leaf with a primary function of conducting photosynthesis. Trophophyll could also be a microphyll that takes part in photosynthesis. Trophophylls are not usually specialized or modified to perform other functions. In Lycophytes, Trophophylls are sterile leaves that are not associated with sporangia. Even though Trophophylls undergo photosynthesis, they do not bear any spores. Photosynthesis leads to the production of sugars. Trophophylls are very similar to sporophylls in that they are both green, photosynthesize and bear some similarities in their anatomy. Despite the similarities, there are also some differences including their modifications.

Trullate

Trullate refers to the shape of a simple leaf with the shape of a trowel blade. The Trullate can be said to be inversely kite-shaped.

Trunk

Trunk is a term used in botany to refer to the stem and main wooden axis of a tree. The trunk is also called bole. The presence of trunks in trees is very useful in the identification of trees. Depending on the species of plants, trunks are marked differently from the bottom to the top. Trunks are present in true woody plants such as monocots and palms. They have different internal physiology. For trees, thickening occurs gradually over the course of time. Thickening is a result of the formation of secondary growth (Novello et al., 2020). Trunks are usually protected by a protective layer known as bark. This is because trunks are very vulnerable to damage including sunburns. Trunks are made up of five main parts namely the heartwood, sapwood, cambium, inner bark and bark. From the outside of the trunk is the bark which is followed by the inner bark. The inner bark is made up of phloem which is very useful in transporting nutrients from the roots to shoots. The inner bark is followed by cambium which is a thin layer of undifferentiated cells. The cells in the cambium divide to replenish phloem cells and xylem cells.

The cambium also contains the growth meristems of the trunk. The cambium is followed by sapwood that carries the living xylem cells. The xylem cells are actively involved in the transportation of water throughout the tree and storage of starch. At the center of the trunk is the heartwood. The heartwood is made up of old xylem cells. These xylem cells are filled with minerals and resins which keeps the organism from infecting the center of the tree.

Truss

Truss is a compact cluster of fruits or flowers that arise from one center. Trusses tend to emerge at or near the point where the primary stem meets a leaf stem. Trusses can produce multiple small fruits or large fruits depending on the variety of plants. In tomatoes, the truss produces yellow flowers. From the center of the yellow flowers, small green tomatoes will appear. Rhododendrons bear flowers on trusses. In the cultivation of flowers, there are cases where the flowers are pruned. Pruning shears are used to snip trusses at their bases. Pruning mostly involves the removal of old flowers.

Tuber

A tuber is a type of specialized vegetative underground storage organ in certain plants. Tubers accumulate water and food. They also prevent the death of plants by drought, fire or other circumstances. Tubers play a vital role in plant perennation and ensuring regrowth during the next season. Thickened rhizomes are formed from stem tubers. Yams and potatoes are some of the species with stem tubers while dahlias, cassava and sweet potatoes are root tubers. Stolons are also formed from stem tubers. Shoots are produced from the top side of the stem tuber. The shoots then grow into leaves and stems (Mercuri et al., 2002). The bottom part of stem tubers grows to produce roots. In most cases, new tubers are attached to a parent tuber. During autumn, new tubers remain while the plant dies. The new tubers have a dominant bud that regrows during spring to form new shoots that produce the new stems and leaves. Other plants have smaller tubers called tubercules. The tubercules play the role of seeds. In some plants, the stem tubers are long-lived while in others they are short-lived. For potatoes, the stem tubers are formed from stolons which thicken and form storage organs.

Root tubers are also called storage roots. They are modified lateral roots that have been enlarged to function as storage organs. The entire root or parts of the roots could be enlarged. Root tubers are also penetrative organs and play a vital role in ensuring the survival of the plant. In plants like sweet potatoes, their root tubers are formed from a massive enlargement of secondary roots. Root tubers do not have any nodes or internodes. Crown tissues are located at the proximal end of root tubers. Crown tissues produce buds that grow into foliage and stems.

Tubercle

A tubercle is a round nodule, wart outgrowth or small eminence found on the internal and external organs of a plant. The meaning of a tubercle may vary depending on the family of plants. For plants like cacti and orchids, it means a warty outgrowth, small eminence or round module that is found on the lip. Tubercles are sometimes called podaria. For plants in the pea family, tubercle refers to the wart-like excrescences found on the roots. For fungi, the term tubercle is used in mycology to refer to a mass of hyphae from which mushrooms are made (Matyjaszczyk & Śmiechowska, 2019).

Tuberculate

Tuberculate also called tubercular or tuberculated means a surface that has been covered by tubercles. This means a part of a plant that has been covered by small rounded warts, nodules or protuberances. The term was formed from the Latin word "tuberculum". The Latin term means affected or covered with small rounded warts. In some cases, tuberculate is about the same as verrucose. Cactus is an example of a plant with tuberculates. The stem of the cactus is covered with podaria or tubercules.

Tuberoid

A tuberoid is an alternative name for the underground storage organ formed by the swelling of a root. Tuberoids occur in most species of orchids.

Tuberous

Tuberous is a term used in botany meaning to resemble a tuber or that which produces tubers. There are serval plants that bear tuberous roots. They include taro, sweet potato, irises, peonies, daylilies, dahlia and asparagus. For such plants, there are swollen roots at the base of the plant. Tuberous plants are mostly planted in spring. They grow and produce normal roots which spread out. As the plant grows, its shoots produce carbohydrates which are directed towards the roots where they are stored (Mandel, 2019). Carbohydrates are stored in the roots as starch. The swollen roots do not have nodes and in most cases, they develop into adventitious buds. This enables the propagation of most plants. For instance, in sweet potatoes, propagation occurs by slipping the tuberous root. Tuberous roots have the ability to produce new plants. In most cases the tuberous roots grow in a tangled mass of stems and roots. A new plant arises from the cluster of storage roots. There growth rates tend to be faster compared to other means of propagation.

Tubular

Tubular is a term used in botany to refer to having a form of a cylinder or a tube. There are several plants whose flowers have a tubular form. The Virginia bluebell is a plant with purple flowers that are tubular in shape. Others include firecracker plants, flower tobacco, tall verbena, red-hot poker and crocosmia. All the flowers of these plants attract the humming bird as its beak has been modified to enable it draw nectar.

Tufted

Tufted in botany refers to plants growing in small dense clumps or tufts. It also refers to plant life as they do not possess the ability to locomote. It also refers to plants that grow together in groups that are not close together.

Tunic

Tunic is the outer covering of some corns and bulbs. Also called tunica, tunic is a technical term in botany. It is a membranous structure that covers or lines organs in plants. Plant organs that are covered with a tunic are said to be tunicate. Therefore a bulb having a tunic will be called a tunicated bulb (Lindström et al., 2013). In some angiosperms, the apical meristem is covered by a tunic which is composed of a layer of cells. The tunic plays the role of ensuring growth and development. In corns and bulbs, the tunic plays a protective role. Its tunic is composed of protective outer coverings of dead materials. Initially, the spermoderm or testa is at times called tunic. The peridium in fungi is also referred to as tunic.

Tunicate

Tunicate is a term used to refer to organs covered with a tunic or having concentric coats. The tunic in tunicate plants performs different functions depending on the type of plant. Bulbs are generally known to be tunicate.

Alliums, grape hyacinths, hyacinths, daffodils and tulips are examples of tunicate bulbs. In bulbs, the tunic is useful in preventing the bulb from drying. In angiosperms, the tunic helps in ensuring development and growth.

Turbinate

Turbinate refers to fruits or roots shaped like a beetroot or spinning top.

Turgid

Turgid is a term used to refer to plant cells that are swollen with liquid. The phenomenon of a plant cell is turgid is called turgidity. Turgidity explains how plants cells are able to stand upright even though they do not have a skeletal structural framework. Cell walls are able to remain turgid because of the presence of a cell wall. The osmoregulatory function of the vacuole is useful in ensuring plants cells are turgid. The cell wall also protects the cell from cell lysis.

Tussock

Tussock refers to a dense tuft of vegetation that is separated from neighboring tussocks. For grasses, a tussock is mostly found in the family Poaceae. Tussock grasses grow as singular plants in hummocks, tufts or clumps.

Two-ranked

Two-ranked is a term that refers to leaf arrangement where the leaves are arranged in two rows in the same plane. Leaves are usually arranged on opposite or alternate sides of the branch. The two-ranked arrangement in leaves is also called the phyllotaxis or phyllotaxy. If there are several leaves arising, the leaves may be whorled at the same node on a stem (Luan et al., 2020).

Type

Type is an item to which the name of a taxon is permanently attached. For instance, type could be a designated representative of a plant name. Type is very important in determining the priority of names available for a particular taxon.

Type genus

Type genus in nomenclature is a single genus on which a taxonomic family is based. In botanical nomenclature the type genus has no status in the ICN however, the type specimens are used when ranking up to family. Types tend to be optional in higher ranks.

U

Umbel

An umbel is a type of flowering in botany that contains a set of small pedicels or fairly brief flower stalks that emerge from a center element, similar to umbrella ribs. The term was invented in botanical terminology in the 1590s, derived from the Latin umbrella, which means "parasol, sunshade." The layout can range from flat-topped to nearly spherical (Lambers et al., 2008). Umbels can be simplistic or complicated in nature. Umbellules or umbellets are the auxiliary umbels of compound umbels. An umbellule is a tiny umbel. Umbellate, or alternatively subumbellate, describes the set of the inflorescence in umbels (almost umbellate).

Each pedicel in an umbel begins at almost the same location at the end of the peduncle, producing the illusion of an umbrella-like form, as in wax flowers (Hoya).

Figure 178: Compound umbels of common hogweed (*Heracleum sphondylium*).

Source: https://www.britannica.com/science/umbel

Umbo

A mushroom cap's umbo is an elevated region in the middle. Umbonate caps are those that have this property. Sharply pointed umbos are referred to as acute, whereas rounder umbos are referred to as broadly umbonate. The umbo is cuspidate if it is elongated, and mammilate or papillate if it is finely outlined but not elongated (similar to the shape of a mammalian areola).

Figure 179: *Cantharellula umbonata* has an umbo.

Source: https://en.wikipedia.org/wiki/Umbo_(mycology)#/media/File:Cantharellula_umbonata_59168.jpg

A protuberance, such as those found on the extremities of the plates of certain pine cones.

Some pines have prominent umbos which can be thorny.

Umbonate

The dictionary defines umbonate as a tiny hump emerging from the center of the cap of certain mushrooms. Another definition of umbonate is a hooked prominence at the tip of each side of the shell of a bivalve mollusk, although this falls outside the scope of plant and floral research (Klassen, 2001).

Unciform

- bent like a hook; hooked
- hooked; bent at the end like a hook

Uncinate

- hooked; bent at the end like a hook
- hooked at the apex

Undershrub

Subshrub; a small, usually sparsely branched woody shrub less than 1 m high.

A short woody plant known as a subshrub, undershrub, or dwarf shrub. A similar phrase is prostrate shrub. The terms "bush" and "subshrub" are considered synonymous.

The definition of a subshrub is not easily distinguished from that of a shrub because the criteria are of degree rather than kind; instances of reasons for characterizing plants as subshrubs entail ground-hugging stems or a low growth habit (Katova, 2007). Subshrubs may be primarily herbaceous, with overwintering perennial woody vegetation growing considerably slower than deciduous summer growth. Some plants classified as subshrubs are only slightly woody and last only a few years; others, like *Oldenburgia paradoxa*, remain indefinitely, rooted in stony fissures.

Small, low shrubs such as lavender, periwinkle, and thyme, and many members of the family Ericaceae, such as cranberries and small species of Erica, are often classed as subshrubs.

Understory

In forestry and ecology, the understory, often referred to as underbrush or undergrowth, refers to plant life that grows beneath the forest canopy but above the forest floor. Because only a small amount of light breaches the canopy, most understory vegetation is shade-tolerant. The understory is often made up of trees that have been stunted due to a lack of light, other tiny trees with low light demands, saplings, shrubs, vines, and undergrowth. Small trees like holly and dogwood thrive in the understory.

Many understory plants in temperate deciduous forests begin to grow earlier in the year than canopy trees in order to take advantage of the expansive abundance of light at that time of year.

A breach in the canopy left by a tree's death motivates potential emergent trees to compete for space by growing upwards to fill the gap. These trees often have straight stems and a small number of lower branches (Jeger et al.,

2007). Simultaneously, the forest floor's shrubs, undergrowth, and trees and plants get denser. The understory is more humid than the canopy, and the covered ground does not fluctuate in temperatures as much as open terrain. This promotes the growth of ferns, mosses, and fungus and promotes nutrient recycling, resulting in suitable habitats for many flora and fauna.

The understory is the underlying level of vegetation in a forest or wooded area, consisting primarily of trees and shrubs spreading between the tree tops and the forest floor. Plants in the understory include the young and juveniles of canopy trees, as well as specialty understory shrubs and plants. Young canopy trees can often be found suppressed in the understory for years until a break in the forest overstory allows them to grow into the canopy. Understory shrubs, on the other hand, finish their life cycles in the shadow of the forest canopy. Many smaller tree species, including the dogwood and holly, hardly grow tall and are typically found in the understory.

The canopy of a tropical forest is typically 10m thick and intercepts approximately 95% of the sunlight. As a result, the understory absorbs less direct light than vegetation in the canopy, and the light that does penetrate is deficient in wavelengths that are most useful for photosynthesis. As a result, understory plants must be shade tolerant—that is, they must be able to photosynthesize effectively with the light that does reach their leaves. They can frequently employ wavelengths that crown plants cannot.

Figure 180: Lesser celandine (*Ranunculus ficaria*) on the forest floor in spring.

Source: https://en.wikipedia.org/wiki/Understory#/media/File:Celandine_Sward.JPG

Undulate

Undulate: having a wavy margin.

Undulate is a word that implies to flow in a wave-like way. If a sound raises and drops in pitch or volume in a wave-like pattern, the sound is said to be undulating. The rescuers' shouts echoed through the jungle as they searched for the missing kid (Hirschberg, 2001).

Undulate is derived from the Latin term undula (unda means "wave"). So every activity denoted by the verb undulate has a waving quality to it. It is said to undulate if anything moves in a wavy manner, for example. The wind may cause prairie vegetation to undulate, whereas music may stimulate hips to undulate.

Uniflor

Having a single flower (uniflory). Compare pauciflor (few) and pluriflor (many).

Monotropa uniflora, also referred to as the ghost plant, ghost pipe, or Indian pipe, is a herbaceous perennial plant indigenous to temperate Asia, North America, and South America, with huge gaps in between. The plant can be totally waxy white, but it also contains black specks or pale pink colors. Rare varieties may be dark red in color.

M. uniflora flowers are visited by a variety of bee and fly varieties, the most common of which are bumblebees. Bumblebees play a crucial role in pollen distribution for the plant.

The three communities of the plant are genetically diverse from one another, according to DNA study. Furthermore, the American cousin populations appear to be more closely connected to one another than to the Asian population.

M. uniflora possesses 48 chromosomes.

Unilocular

A locule, often spelled loculus, is a tiny chamber or pocket within an organ or component of an entity (animal, plant, or fungus).

In flowering plants, a locule is a compartment inside an ovary that contains the flower and fruits. Fruits are classed as uni-locular (unilocular), bi-locular, tri-locular, or multi-locular based on the number of locules in the ovary. A gynoecium's locule count may be equivalent to or fewer than the amount of carpels. The ovules or seeds are housed in the locules (Hadwiger, 2013).

The phrase can also apply to pollen-containing chambers within anthers.

Lociles are compartments within the hymenium where perithecia grow in Ascomycete fungi.

Figure 181: A bi-locular and a multi-locular tomato fruit.

Source: https://en.wikipedia.org/wiki/Locule

Uniserial/ Uniseriate

A uniseriate is a single whorl of the accessory (non-essential) floral part. It is a term used to describe flowers that are unbranched or otherwise organized in just one row or series.

Unisexual

Gonochorism is a sexual system in biology in which there are exactly two sexes and each single entity has been either male or female. The term gonochorism is most commonly used in reference to animal species, the overwhelming bulk of which are gonochoric (Guo & Liu, 2007).

Gonochorism differs from simultaneous hermaphroditism, however, it can be difficult to establish whether a species is gonochoric or hermaphroditic. Individuals of gonochoric species, on the other hand, stay either male or female all their lives. Gonochoric species are those that reproduce via parthenogenesis so don't have males.

The phrase comes from Greek. The term gonochorism, though, was later was derived from the German word gonochorismus.

Unisexualism or gonochory are other terms for gonochorism.

Plants with single-sex individuals are commonly referred to as dioecious rather than gonochoric. Individual flowers of flowering plants, for example, may be hermaphroditic (containing both stamens and ovaries) or dioecious (unisexual), with either no male reproductive organs or no female reproductive organs (Franklin et al., 2013). Some blooming plants with unisexual flowers also produce hermaphrodite flowers, and the three forms can coexist on the same or distinct plants. Hermaphrodite, dioecious, trioecious, polygamomonoecious, polygamodioecious, andromonoecious, or gynomonoecious plant species are thus possible.

Holly and kiwifruit are two examples of species having gonochoric or dioecious pollination. The pollenizer in these plants is the male plant that delivers the pollen.

Figure 182: Kiwifruit by species A = A. arguta, C = A. chinensis, D = A. deliciosa, E = A. eriantha, I = A. indochinensis, P = A. polygama, S = A. setosa.

Source: https://en.wikipedia.org/wiki/Kiwifruit#/media/File:Actinidia_fruits.jpg

Unitegmic

An integument is a layer of cells that protects the ovule. Gymnosperms have one integument (unitegmic), whereas angiosperms have two integuments (bitegmic). The evolutionary genesis of the innermost integument, which is essential for the generation of ovules from megasporangia, has been hypothesized as sterile telomes enclosing a megasporangium (Arruda et al., 2015). Elkinsia is a preovulatory taxon with a lobed formation attached to the bottom third of the megasporangium, the lobes spreading upwards in a circle around the megasporangium. This could have resulted in an integument as a result of lobe fusion and fusion across the structure and the megasporangium.

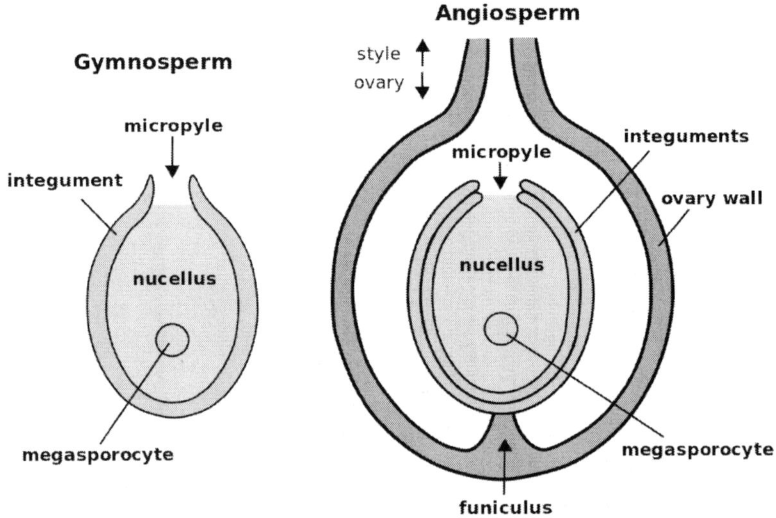

Figure 183: Plant ovules: Gymnosperm ovule on left, angiosperm ovule (inside ovary) on right.

Source: https://en.wikipedia.org/wiki/Ovule#/media/File:Ovule-Gymno-Angio-en.svg

For some time, the genesis of the secondary or exterior integument has been a source of controversy. A few angiosperms have vascular tissue in their outer integument, which indicates that the external layer is morphologically abaxial. This shows that cupules like those generated by the Caytoniales or Glossopteridales could well have developed into angiosperms' exterior integument (Foxcroft et al., 2017).

When the ovule develops following fertilization, the integuments grow into the seed coat.

Urceolate

Urceolate is defined as "formed like an urn."

An urceolate corolla is similar to a campanulate corolla in overall design, and rather than being bell-shaped, it is fashioned like an urn and is extended in the center, tapering towards its opening. Urceolate corollas are seen on several species in the heath family (Ericaceae). Vaccinium corymbosum (highbush blueberry), which naturally occurs in bogs and marshy forests and is the most often farmed blueberry for commercial purposes, is such an example.

Usu

Usually abbreviated is usu

Utricle

An achene is a kind of dry fruit borne by many flowering plant species. Achenes are monocarpellate (made up of only one carpel) and they do not open at maturity. Achenes have a single seed that almost completely fills the pericarp but does not cling to it (Fahad et al., 2017). In many species, the "seed" is actually an achene, which is a fruit that contains the seed. The seed-like look is caused by the hardening of the fruit wall (pericarp), which completely covers the solitary seed so tightly that it seems to be a seed coat.

A utricle is similar to an achene, except that the fruit is pouched or inflated.

Utricles are air-filled sacs found in certain seaweeds and can be found in aquatic plants.

V

Vallecular Canal

A cortex is the outer layer of a plant stem or root that is under the epidermis but outside of the vascular bundles. It is also the surface layer of some lichens' nonfruiting parts of their bodies.

The cortex is largely made up of massive, thin-walled parenchyma cells from the ground tissue system, with minimal to no structural distinction. Collenchyma cells are outer cortical cells that have irregularly thickened cell walls. Chloroplasts may be seen in some of the outermost cortical cells. Cork is formed by layers of cells formed by the cortex. It is in charge of transporting resources into the inner cylinder of the root via diffusion and may also be employed for starch storage (Elmer & White, 2018).

The endodermis is the cortex's deepest layer.

The vallecular canal is one of the vast intercellular passageways of the cortical parenchyma that alternates with the vascular bundles in the branches of Equisetum plants.

Valvate

Aestivation, also known as estivation, is the organization of the components of a flower inside a flower bud prior to its opening. Aestivation is also known as praefoliation or prefoliation, however, these names can also apply to vernation, which is the positioning of leaves inside a vegetative bud.

Aestivation can be a useful taxonomic distinguishing feature; for example, Malvaceae flower buds have valvate sepals, besides the genera Fremontodendron and Chiranthodendron, that have occasionally been misidentified as a consequence.

The words used to indicate aestivation and leaf vernation are interchangeable.

Valvate aestivation occurs when the margins of neighboring petals or sepals touch without overlapping.

Figure 184: Lilac (*Syringa vulgaris*), valvate aestivation.

Source: https://en.wikipedia.org/wiki/Aestivation_ (botany)

Valve

Plants have devised a variety of creative methods for assuring the distribution of their seeds when they reach maturity. Berries, for example, contain tiny seeds inside every nodule of the fruit, such as blackberries and raspberries. When animals ingest the berries, they also swallow the seeds, which proliferate when the seeds transit through the gastrointestinal tract of the animal that ate them.

Other plants have evolved various methods for assuring seed dissemination after fruit development. They have evolved fruit forms and mechanisms that guarantee the fruit's shell splits open during maturation, enabling the seeds to fall out. In this case, each segment of the split pericarp is referred to as a valve (De Waele & Elsen, 2007).

Consider the humble bean. They grow in pods on the plant. As the pods develop, they split into two distinct halves, commonly divided by a central groove containing a "string." The fruit begins to dry as it grows. Once fully dried, the pod splits into two valves, allowing the beans (seeds) to fall out and sprout on the soil, or be ingested by animals and spread to other locations by waste dispersal.

In the world of botany, the various bean variations we currently enjoy are only a few examples of valves. There are numerous others, including orchids (vanilla beans), a variety of nuts, and even some legumes.

Var.

An abbreviation of varietas.

Variant

A plant or grouping of plants that differs in some way from the features associated with a specific taxon.

The word variant or "genetic variant" is used in microbiology and virology to define a subtype of a microbe that is genetically unique from a primary strain but not substantially dissimilar to be called a distinct strain. In botany, a comparable difference is established between several cultivated variations of a plant species, known as cultivars (Chen & Yamaguchi, 2002).

According to one report from 2013, "there is no universally agreed definition for the phrases 'strain,' 'variant,' and 'isolate' in the virology field, and most virologists merely copy the application of terms from others."

In the context of the Variant of Concern 202012/01 version of the SARS-CoV-2 virus, the US Centers for Disease Control and Prevention (CDC) website states, "For the time being in the context of this variant, the [terms "variant," "strain," and lineage"] are generally being used interchangeably by the scientific community."

Variegated

The emergence of different colored zones in the leaves, stems, and fruit of plants is known as variegation.

Variegated species are occasionally found in the substory of tropical rainforests, and this environment is the origin of a variety of variegated home plants.

The phrase is also used to describe color zonation in plants, minerals, and animal skin, fur, feathers, or scales.

Scientists have examined virus infection as a cause of variegation, such as in the instance of the Tobacco Mosaic Virus.

Many gardeners enjoy the variegated effect and have purposefully attempted to create it for aesthetic purposes.

There are several gardening publications on the subject, as well as gardening groups that specialize in the subject.

Because the variegation is caused by the presence of two types of plant tissue, the plant must be propagated using a vegetative manner that retains both kinds of cells in respect to each other (Bourgaud et al., 2001).

Stem clippings, bud and stem grafting, and other reproduction techniques that result in development from leaf axil buds, in general, will sustain variegation. Cuttings with perfect variegation may be hard to spread, if not impossible.

Because the new stem tissue is produced from a specific tissue type inside the root, root cuttings frequently do not sustain variegation.

The concealing of green pigment by other pigments, such as anthocyanins, is a typical source of variegation. This frequently extends throughout the entire leaf, turning it scarlet or purplish. Some plants, however, have constant zonal markings, such as clovers, bromeliads, and certain Pelargonium and Oxalis species. On others, like as frequently grown Coleus, variegation can differ tremendously within a group.

Virus infections may result in striping on the leaf surface. The design is frequently indicative of an infection. Mosaic viruses, which cause a mosaic-like effect on the leaf surface, and citrus variegation viruses are two examples (CVV).

Varietas

In popular parlance, variation is sometimes shortened as var.

A taxonomic level lower than species and between subspecies and forms.

Vascular

Vascular tissue is a complex conducting tissue found in vascular plants that is made up of more than one cell type. The xylem and phloem are the fundamental components of vascular tissue. Essentially, these two types of cells transfer fluid and nutrients. The vascular cambium and the cork cambium are both meristems connected with vascular tissue. The vascular tissue network of a plant is made up of all of the vascular tissues inside that plant.

Vascular tissue cells are typically long and thin. Because the xylem and phloem participate in the transport of water, minerals, and nutrition throughout the plant, it is not unexpected that they have the shape of pipes. Individual phloem cells are joined end-to-end, precisely like pipe sections (Chandler & Sanchez, 2012).

New vascular tissue changes at the plant's developing tips as it expands. The new tissue is aligned with the existing vascular tissue, allowing it to remain connected all through the plant. Plant vascular tissue is organized into long, distinct strands known as vascular bundles. Such bundles contain xylem and phloem, as well as sustaining and protecting cells. In shoots and roots, the xylem is normally found closer to the inside of the stem, while the phloem is found closer to the exterior. Some Asterales dicots' stems may also have phloem positioned internally from the xylem.

The vascular cambium is a meristem located between the xylem and phloem. This tissue splits into cells that will produce more xylem and phloem.

This growth expands the plant's girth instead of its height. The plant will keep developing stronger as long as the vascular cambium keeps producing new cells. The vascular cambium promotes the expansion of vascular tissue that generates woody growth in vegetation that makes wood.

Vascular Bundle

In vascular plants, a vascular bundle is a component of the transportation network. The transport takes place in the stem, which comes in two varieties: xylem and phloem. Both of these tissues are found in a vascular bundle, which also contains supportive and protecting tissues. The cambium is a tissue that exists between the xylem and the phloem (Clemens et al., 2013).

The xylem is normally located near the axis, while the phloem is located further from the axis. This signifies that the xylem is nearer the center of the stem or root, although the phloem is nearer to the outside. The adaxial side of a leaf is normally on the upper side, while the abaxial surface is on the lower side.

The sugars generated by the plant with the help of sunlight are transferred by the phloem, which is located closer to the ground. Aphids and leafhoppers feed on these sugars by entering the phloem. That's why aphids and leafhoppers prefer to live on the underside of leaves instead of the top. The location of vascular bundles in relation to one another can vary greatly.

Vasculum

Botanists utilize a vasculum, also known as a botanical box, to preserve field samples intact for transfer. The major function of the vasculum is to convey plants without damaging them and to keep the environment pleasant and damp.

Vascula are cylinders that are normally constructed of tinned and often coated iron, however wooden counterparts exist. The box was held horizontally on a sling so that the plant specimens lay flat, and it was covered with damp fabric. Originally, British and American vascula were flat with a single room, whereas continental counterparts were more tubular and typically longer, with two separate sections (Zehnder et al., 2007).

Accessibility to the interior is provided by one (or sometimes two) huge lids on the side, letting plants to be placed and removed without bending or twisting them needlessly. This is especially crucial with wildflowers, which are frequently fragile.

Some early-twentieth-century specimens are fashioned of sheet aluminum instead of tin, but generally, adhere to the 19th-century form. The outside is typically left plain or lacquered green.

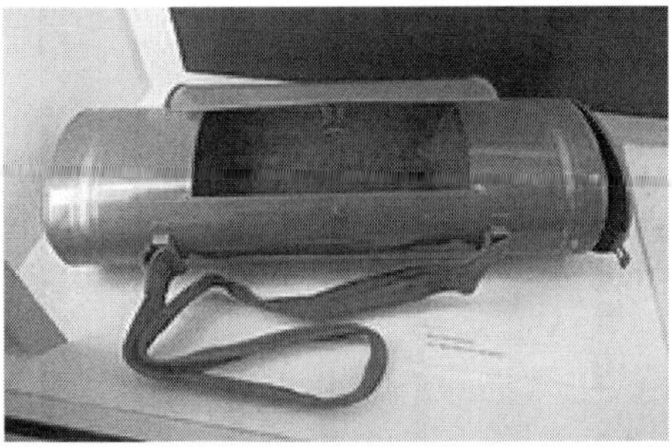

Figure 185: Vasculum.

Source: https://en.wikipedia.org/wiki/Vasculum

Vein

Veins (also known as nerves) are one of the more apparent leaf qualities or features. The veins in a leaf reflect the organ's vascular structure, reaching into the leaf through the petiole and transporting nutrients and water between the leaf and stem, and they play an important role in the preservation of leaf hydration status and photosynthetic potential. They also help with the mechanical stability of the leaf. Although some vascular plants have only a single vein inside the lamina of the leaf, in most this vasculature splits (ramifies) as per a variety of patterns (venation) and forms cylindrical bundles, usually resting in the median plane of the mesophyll, between the two layers of the epidermis (Yamada et al., 2003).

This pattern is frequently taxonomic, with angiosperms having two primary kinds, parallel and reticulate. Generally, monocots have parallel venation, whereas eudicots and magnoliids ("dicots") have reticulate venation, however, there are several exceptions.

Primary sometimes called first-order veins are those that reach the leaf from the petiole. Secondary veins or second-order veins originate from the primary veins. Both primary and secondary veins are classified as major or lower-order veins, while some authors add third-order veins. Every branching after that is numbered sequentially, and these are the upper order veins, with each spreading connected with a smaller vein width.

In comparison, leaves with reticulate venation have a single (occasionally several) principal vein in the center of the leaf, known as the midrib or costa, which is continuous with the vasculature of the petiole more proximally. The midrib then divides into several smaller secondary veins, sometimes referred to as second-order veins, which stretch toward the leaf margins. These are frequently terminated at the margin by a hydathode, a secretory organ. Smaller veins, known as tertiary or third-order (or higher-order) veins, branch from the secondary veins, creating a dense reticulate structure. Areoles are regions or islands of mesophyll that exist among higher-order veins.

Veinlet

A small vein; the ultimate (visible) division of a vein.

Areolation occurs when some of the smaller veins (veinlets) terminate in the areoles. These small veins serve as exchange points between the mesophyll and the plant's circulatory system (Yakhin et al., 2017). Thus, minor veins collect photosynthetic products (photosynthate) from the tissue where it occurs, whereas major veins transfer it out of the leaf. Water is being moved in the other direction at the same time.

Velamen

Velamen, also known as velamen radicum, is a spongy, multilayered epidermis found on the roots of several epiphytic or semi-epiphytic plants such as orchids and Clivia species.

An orchid's velamen is the white or gray coating of its aerial roots It has several cell layers and is capable of absorbing ambient water and mineral nutrients, but its primary role may be to shield the underlying cells from UV radiation.

Orchid roots are frequently connected with symbiotic fungi or bacteria, which may absorb nutrients from the atmosphere.

This feature enables the orchid to exist in environments that yield a reproductive or developmental value, such as more exposure or less pressure from other plant species.

The velamen also provides a mechanical role, insulating the root cortex's vascular tissues from transpirational water loss and, in many situations, attaching the plant to the ground.

A normal orchid root has a stele that is quite modest in diameter. It is encircled by a cortex, which is further covered by an extremely specialized exodermis, the majority of which do not possess protoplasm at maturity. Nevertheless, a few cells are alive and enable water to move through them (Yakhin et al., 2017).

The exodermis is bordered by velamen, which is made up of one to many layers of cells that can generate root hair under the right conditions.

The velamen develops from the root tip through the division of a specific tissue. Velamen's dead cells absorb light, lending it a grey look at the tips, whereupon chlorophyll is apparent. When the dead cells retain moisture, they become translucent, and the entire velamen tissue turns green.

Velutinous

Having a soft, velvety surface, as certain plants.

Velvety

Densely covered with fine, short, soft, erect hairs.

Venation

Originating with Ettingshausen and many other descriptive phrases, a number of distinct classification methods of the configurations of leaf veins (venation or veination) have been documented, and the nomenclature has been regarded as "formidable." Among the most frequent is the Hickey system, which was originally established for "dicotyledons" and employs a number of Ettingshausen's terminology borrowed from Greek.

Ventral

The word ventral comes from the Latin word ventre, which means "belly." The inverse of dorsal. Because the phrase historically referred to animals rather than plants, its application in botany varies depending on context and provenance (Wisler & Norris, 2005). In general, "ventral" alludes to "the belly or lower region," however in botanical terminology, such concepts are not always well defined and can be contradictory. As an example:

- When speaking to a lateral part of an erect plant, it should be towards the axis (adaxial).
- Any section of an erect plant that faces the substrate, such as the lower surface of a more or less horizontal leaf (abaxial)
- In a prostrate or climbing plant, facing the substrate.

Vernation

Vernation is the creation of new leaves or fronds. It is derived from the word vernal, which means spring, because that is when leaves sprout in temperate countries. This is the configuration of leaves in a bud in botany.

New leaves of pine species are small and wrapped in sheaths. Each leaf bundle has 2 to 5 needles. All of the leaves on a single branch increase in length at the same time. New leaves of cabbage species are wrapped over and concealed by the preceding leaf.

The process through which a fern frond arises is known as circinate vernation. The fern frond is tightly curled as it grows, protecting the sensitive growing point of the frond within coils.

Figure 186: Cycad leaves are produced by involute vernation.

Source: https://en.wikipedia.org/wiki/Vernation#/media/File:Involute_vernation.jpg

The folding of one border of the leaf's blade over the other is the action of convolute vernation. Its folding sytems gives the developing leaf the appearance of a tube.

Involute vernation occurs when both edges on opposite ends of the leaf roll up more toward the upper (axial) surface of the leaf, generating two tubes that may connect at the leaf's midrib (Weitemier et al., 2014).

The inverse of involute vernation is revolute vernation, in which the leaf margins roll up towards the beneath (abaxial) surface of the leaf.

Vernonioid

A type with sweeping hairs borne on abaxial surfaces of style branches.

Verruciform

Resembling a wart; denoting wartlike elevations.

Verrucose

Resembling a wart; denoting wartlike elevations.

Verruculose

Having warts.

Versatile

The categorization of stamens is determined by the attachment of the filament to anthers. The adaptable anther is one that is linked to the filament at a spot on the connective, allowing it to swing freely. Grass exhibits this condition.

Verticillate

A whorl or verticil in botany is a pattern of leaves, sepals, petals, stamens, or carpels radiating from a single point and surrounding or wrapping around the stem or column. A leaf whorl is made up of at least three parts; a whorl is not made up of two opposed leaves.

Grouped in one or more whorls, that is, several comparable elements arranged at the same location on the axis, such as a leaf layout.

Verticillaster

A Lamiaceae type; a pseudoverticillate inflorescence in which pseudo-whorls are created by pairs of opposite cymes (Wilkinson et al., 2012).

This indicates that they are numerous flowers tightly packed together to make an effective structural unit, although they look to be whorled (verticillate) but are not however they form in a way in which the main axis and all lateral branches end in a flower

Vesicular

Bladder-like; vesciculous, bearing such hairs.

Vessel

A vessel element or vessel member (sometimes known as trachea or xylem vessel) is a specialized cell present in the xylem, plants' water-carrying tissue. Vessel components are common in angiosperms but are lacking in most gymnosperms, including conifers. Vessel components are the primary trait that distinguishes angiosperm "hardwood" from conifer "softwood."

Vestigial

Vestigiality is the retention of genetically inherited features or traits that have shed some or all of their ancestral purpose in a specific species the along evolutionary process. In general, assessing vestigiality must rely on contrast with homologous traits in similar species. Vestigiality emerges through normal evolutionary processes, generally through the loss of function of a trait that is no longer susceptible to positive selection forces when it sheds its usefulness in a dynamic world (Waraich et al., 2011). The trait may be chosen more urgently when its function is clearly damaging, but if the absence of the feature gives no advantage and the existence of the character gives no disadvantage, the characteristic may not be phased out by natural selection and may survive across species.

Vestigial portions of plants include superfluous stipules and carpels, Equisetum leaf reduction, and Fungi paraphyses. Reduced floral display, resulting in smaller and/or paler blooms, is a well-known illustration in plants that multiply without outcrossing, such as selfing or obligate clonal reproduction.

Villous

Has plenty of or coated with long, soft, fine hairs; scruffy with soft hairs

Vine

A vine is any plant that grows with trailing or scandent (climbing) stems, lianas, or runners. They can be used in wicker craft, the term vine can also apply to the stems or runners directly.

In some parts of the world, particularly the British Isles, the term "vine" refers solely to grapevines (Vitis), whereas the phrase "climber" refers to all climbing plants.

Some plants grow as vines all the time, whereas others grow as vines only part of the time. Poison ivy and bittersweet, for example, can develop as low shrubs when no structure is present, but will form vines when support is given.

A vine has a growth form that is built on very lengthy stems. This serves two functions. Instead of investing energy in a lot of supporting tissue, a vine may exploit rock exposures, neighboring plants, or other props for development, allowing the plant to access sunlight with a minimal input of energy (Van Dijk et al., 2021). Plants such as kudzu and Japanese honeysuckle, which are both invasive exotics in portions of North America, have found this to be an extremely successful growth style.

Some tropical vines acquire skototropism, or negative phototropism, and grow away from the light. The vine's growth away from light allows it to grasp a tree trunk, that it can then traverse to reach brighter places.

The vine development pattern may also allow plants to colonize huge regions quickly, even if they are not able to climb high. Periwinkle and ground ivy are examples of this. It's also an adaptation to living in locations with small sections of fertile soil next to exposed areas with greater sunlight but little or no soil. A vine can anchor in the soil but have almost all of its leaves in a brighter, more exposed place, allowing it to benefit from the best of both habitats.

The evolution of a climbing inclination has been identified as a major innovation connected with the evolutionary success and diversity of several plant taxonomic groups.

Virgate

Wand-shaped, twiggy, particularly of erect, straight stems Pertaining to a pileus with radiating ribs or lines in mycology.

Viridiplantae

Viridiplantae (basically "green plants") are a group of eukaryotic organisms with 450,000–500,000 species that serve vital roles across both terrestrial and aquatic environments. They are composed of green algae that are largely aquatic, as well as land plants (embryophytes) that evolved from within them. Green algae have historically been thought to exclude land plants, making them a paraphyletic group (Van Bruggen et al., 2016). However, it is correct to think of terrestrial plants as a type of alga. Some authors are beginning to add embryophytes after discovering that they developed from within green algae. They have cellulose-containing cells and primary chloroplasts generated from endosymbiosis with cyanobacteria, which include chlorophylls a and b but lack phycobilins.

The group has been considered as a kingdom in certain classification systems, under various names such as Viridiplantae, Chlorobionta, or generically Plantae, with the latter broadening the typical plant kingdom to include green algae. Researchers that created a taxonomy for all eukaryotes gave this group the name Chloroplastida, which reflects the presence of primary chloroplasts with green chlorophyll. Scientists discarded the name Viridiplantae since some of the species are not plants in the usual sense. The Viridiplantae are divided into two clades: Chlorophyta and Streptophyta, with the basal Mesostigmatophyceae and Chlorokybophyceae. Viridiplantae, together with Rhodophyta and Glaucophytes, are assumed to be part of a wider group known as Archaeplastida or Primoplantae.

Figure 187: Green algae on rocks at ShihTiPing (giant stone steps) coastal area in Taiwan.

Source: https://en.wikipedia.org/wiki/Viridiplantae#/media/File:Taiwan_2009_East_Coast_ShihTiPing_Giant_Stone_Steps_Algae_FRD_6581.jpg

Viscid

Sticky, with a viscous, syrupy secretion coating it.

Vitta

An oil tube found in the fruit of certain plants.

Viviparous

Vivipary happens in plants when seeds or embryos start developing before they separate from the parent. Some Iridaceae and Agavoideae plants produce cormlets in the axils of their inflorescences. These fall, and if conditions are favorable, they can effectively start a whole season over fallen seeds (Van Breusegem et al., 2001). Similarly, several Crassulaceae, such as Bryophyllum, form and drop plantlets poised to sprout from grooves in their leaves. The creation of embryos using somatic tissues is an example of asexual vegetative reproduction, which is equivalent to cloning.

Vivipary refers to reproduction through embryos, such as branches or bulbils, rather than germinating outside from a fallen, dormant seed, as is common in plants.

A few plants are pseudoviviparous, which means that rather than reproducing with seeds, Monocot grasses can multiply asexually by forming new plantlets in their spikelets.

Seagrass species from the genus Posidonia and alpine meadow-grass, *Poa alpina*, are two examples.

W

Warty

A surface that is covered with little round protuberances, particularly in fruit, leaves, twigs, and bark.

Synonym tubercule

A tubercle (meaning "little tuber," Latin for "lump") is a spherical nodule, small prominence, or warty protrusion occurring on the surface or internal organs of a plant or animal (Arruda et al., 2015).

A tubercle is a wart-like outgrowth that has a slightly varied connotation depending on which plant or animal family it is used to allude to.

It refers to a spherical nodule, tiny prominence, or warty protrusion seen on the lip of various orchids and cacti. They are also referred to as podaria. This is used to describe the wart-like excrescences found on the roots of various species of the pea family.

Watershoot

An upright, strong-growing, or epicormic shoot that emerges from the bottom of a shrub or tree, as opposed to a sucker.

Weed

A weed is a plant that is regarded unsuitable in a specific situation, or "a plant in the wrong location." Plants that are undesirable in human-controlled environments, such as farmlands, gardens, lawns, and parkland, are notable instances. The phrase "weed" has no botanical value because a plant that is a weed in one setting is not a weed in another, and if one species of plant is a profitable crop plant, some other species in the same genus may be a major weed, such as a wild bramble spreading among farmed loganberries. Volunteer crops (plants) are treated the same way as weeds in the following crop.

Many plants that are commonly regarded as weeds are also purposely produced in gardens and other manicured settings, where they are sometimes referred to as helpful weeds. Weed also refers to any plant that develops or reproduces aggressively or is invasive far beyond its natural habitat. Further, the term "weed" is sometimes used to refer to organisms beyond the plant kingdom that can thrive in a variety of conditions and reproduce swiftly; in this sense, it has even been used to people.

In agriculture, weed control is critical. Hand cultivation with farm tools, motorized cultivation with tractors, smothering with mulch or soil solarization, deadly wilting with high heat, scorching, or exposure to chemicals with herbicides are some of the methods used (Bourgaud et al., 2001).

Certain weed classes exhibit responses to harsh conditions. That is, disturbed ecosystems in which soil or natural vegetative growth has been harmed or is repeatedly damaged, and disruptions that give weeds an edge over valuable crops, pastures, or ornamental plants. The characteristics of the ecosystem and its perturbations will influence or even dictate which sorts of weed communities emerge as dominant.

Plants that have adapted to naturally existing disturbed settings, such as shifting soils, river banks and deltas, and sites that have been burned repeatedly, are illustrations of such pioneer species. Because human agricultural methods frequently mirror the natural settings in which weedy species have developed, some weeds are functionally preadapted to thrive and multiply in human-disturbed regions such as farm areas, meadows, roadside ditches, and construction sites. These weedy species frequently have an edge over more attractive crop species because they survive and spread quickly, have seeds that stay in the soil seed bank for many years, or have a limited lifespan with numerous generations in the same growing season.

Perennial weeds, on the other hand, frequently have underground stems that expand beneath the soil surface or, like ground ivy (*Glechoma hederacea*), creeping stems that root and spread out over the ground.

Wild

The wild, or wilderness or wildlands (generally in the plural) are natural areas on the planet that have not been dramatically altered by anthropogenic activities, or any nonurbanized terrain that is not extensively cultivated. The phrase has generally related to terrestrial areas, but marine wilderness is gaining popularity. According to recent maps of wilderness, it spans around one-quarter of the Earth's terrestrial surface, yet it is fast degrading due to human activities. There is even less wildness in the ocean, withwhere only 13.2 percent free of heavy human activity (Chandler & Sanchez, 2012).

Some nations enact legislation to safeguard wilderness regions in order to not only preserve what is currently there, but also to foster and advance natural expression and evolution.

These can be established in preserves, ecological preserves, state forests, national parks, and even urban areas near rivers, gulches, and other undeveloped places. These regions are frequently regarded as critical for the survival of specific species, biodiversity, ecological study, conservation, isolation, and recreation. They may also help to maintain past genetic features and provide habitat for wild vegetation and organisms that would be impossible to replicate in zoos, arboretums, or laboratories.

Figure 188: Hammastunturi Wilderness in Lapland, Finland.

Source: https://en.wikipedia.org/wiki/Wilderness

Whorl

A whorl or verticil in botany is a configuration of leaves, sepals, petals, stamens, or carpels radiating from a single point and surrounding or wrapping around the stem or stalk. A leaf whorl is made up of at least three parts; a whorl is not made up of two opposed leaves.

Other than in plant species with very few internodes and several other genera such as Gallium, Nerium, or Elodea, it is uncommon for leaves to form in whorls (Craigie, 2011). Some trees, such as *Brabejum stellatifolium* and others in the Proteaceae family, have leaf whorls (e.g., in the genus Banksia). Congested internodes within leaf whorls alternate with lengthy internodes across whorls in plants like these.

Most flowers (referred to as cyclic flowers) have four types of whorls:

- The calyx consists of one or more whorls of sepals at the base.
- The corolla consists of one or more whorls of petals above the calyx.

- The androecium consists of one or more whorls of stamens, each with a filament and an anther.
- The gynoecium consists of one or more whorls of carpels, each of which bears an ovary, a style, and a stigma.

A flower is considered to be incomplete or imperfect if it lacks any of these floral characteristics. Not all flowers have whorls since the parts might be spirally organized, as in the Magnoliaceae family.

Figure 189: Leaf whorls on a herbaceous *Lilium michiganense.*

Source: https://en.wikipedia.org/wiki/Whorl_ (botany)

Wing

- A membrane extension of a fruit or seed that helps in dissemination, as on pine seeds.
- A thin flange of tissue that extends beyond the typical outline of a structure, as found on the column of some orchids, stems, and petioles (Clemens et al., 2013).
- One of the two lateral petals of a flower in the Fabaceae subfamily Faboideae, positioned between the adaxial standard (banner) petal and the two abaxial keel petals.

Woolly

Very heavily covered in long, matted or entangled hairs that resemble sheep's wool.

X

Xeromorph/Xerophyte

A xerophyte is a plant species that has evolved to survive in environments with minimal liquid water, including a desert or an ice- or winter weather location in the Alps or the Arctic. Cacti, pineapples, and several Gymnosperm plants are examples of xerophytes.

Xerophytes' physical traits (morphology) and underlying chemical processes (physiology) are differentially tailored to preserve water, which is also frequent during dry seasons. Other species can withstand prolonged periods of intense dryness or desiccation of their cells, throughout which their metabolic activity effectively ceases. Xeromorphic plants have such morphological and physiological adaptations.

Because of their deep-spreading roots and ability to retain water, xerophytes such as cacti can tolerate prolonged periods of drought. Their waxy, prickly leaves prevent dehydration. Its fleshy stems can also hold water.

Plants take water from the ground, which evaporates from their leaves and stems; this is referred to as transpiration. During dry times, a typical mesophytic plant would evaporate fluids faster than the rate of soil water intake, causing wilting and possibly death (Chen & Yamaguchi, 2002).

To survive in such water-stressed environments, xerophytic plants have developed a wide range of specific adaptations. They may utilize water from their own reservoirs, allocate water precisely to locations of new tissue growth, or lose less water to the atmosphere, allowing a greater proportion of water in the soil to be channeled to photosynthesis and growth. Varied plant species have different characteristics and techniques for managing water supplies, which allows them to live.

Figure 190: *Cistus albidus* is a xerophyte which grows in European countries such as France, and Italy and North African countries like Morocco.

Source: https://en.wikipedia.org/wiki/Xerophyte#/media/File:Cistus_albidus_in_Sainte_Lucie_Island.jpg

Xylem

In vascular plants, the xylem is one of two forms of transport tissue, the other being phloem. Xylem's primary job is to move water from the roots to the stems and leaves, but it also distributes nutrition. The term xylem is taken from the Ancient Greek word wood xylon; wood is the most well-known xylem tissue, though it can be observed throughout a plant. Carl Nägeli coined the phrase in 1858 (De Waele & Elsen, 2007).

The lengthy tracheary segments that convey water are the most distinguishing xylem cells. Tracheids and vessel elements differ in shape; vessel elements are shorter and link together to form lengthy tubes known as vessels.

Other cell types found in xylem include parenchyma and fibers.

The lengthy tracheary segments that convey water are the most distinguishing xylem cells. Tracheids and vessel elements differ in shape; vessel elements are shorter and link together to form lengthy tubes known as vessels.

Other cell types found in xylem include parenchyma and fibers.

Xylem can be found in:

- Non-woody plants and non-woody portions of woody plants have vascular bundles.
- In woody plants, the secondary xylem is put down by a meristem termed the vascular cambium.
- As part of a stelar structure that is not split into bundles, as found in many ferns.

The first two groups are not mutually exclusive in transitional stages of plants with secondary growth, albeit a vascular bundle will typically contain only major xylem.

Z

Zonate

Light and dark circular bands or rings, usually found on leaves or flowers.

Zygomorphic

Floral symmetry refers to whether and how a flower, specifically its perianth, may be separated into two or more identical or mirror-image portions.

Flowers, on the other hand, may have no axis of symmetry at all, usually because their pieces are spirally arranged (Elmer & White, 2018).

Zygomorphic flowers can be classified into two mirror-image hemispheres by a single plane, similar to a yoke or a person's face. Orchids and the blooms of most Lamiales members are examples. Many authors favor the terms unilateral symmetry or monosymmetry. Because of the asymmetry, pollen can be implanted in specific regions on pollinating insects, which can result in the emergence of new species.

Figure 191: *Satyrium carneum*. Ground orchid with typical zygomorphic floral anatomy.

Source: https://en.wikipedia.org/wiki/Floral_symmetry#Zygomorphic

Zygomorphic flowers are in the minority both worldwide and within individual biomes. Plants with zygomorphic flowers have fewer visitor species than those with actinomorphic blooms. Plant sub-networks with zygomorphic flowers have higher connectance, greater asymmetry, and reduced coextinction robustness for both the plants and the visitor species. Plant taxa with zygomorphic blooms may be more critically endangered as pollinators dwindle.

Zygote

Eukaryotes are organisms whose cells have a nucleus surrounded by a nuclear envelope. Eukaryotes are members of the domain Eukaryota or Eukarya. One of the three kingdoms of life is Eukaryota; the other two domains are bacteria and archaea (prokaryotes).

A zygote is a eukaryotic cell generated by the fertilization of two gametes. The zygote's genome is a synthesis of the DNA from each gamete and carries all of the genetic information required to generate a new individual organism (Fahad et al., 2017).

The zygote is the initial developmental stage of multicellular organisms. A zygote is generated when an egg cell is fertilized by a sperm cell in humans and most other anisogamous organisms. Mitosis allows the zygote in single-celled organisms to split asexually and create identical progeny.

In the late 1800s, German zoologists Oscar and Richard Hertwig made some of the earliest discoveries of animal zygote development.

In plants, having more than two alleles in a zygote is not necessarily a bad thing. Plants frequently have multiple copies of a gene with distinct alleles. Numerous plants do not go through meiosis before reproducing by chance, and as a result, they are prone to amassing many copies of a chromosome as generations pass (Foxcroft et al., 2017). The zygote formed by unintentional gametes has the potential to produce unique plant types. Numerous commercial agricultural crops are the result of polyploidy influencing the plant in some way, either growing it bigger or rendering its fruits sterile, requiring consumers to buy more. A zygote is still created when two gametes fuse in plants, despite these changes. When pollen fertilizes an egg inside of ovules in the flower in many higher plants, a zygote is formed.

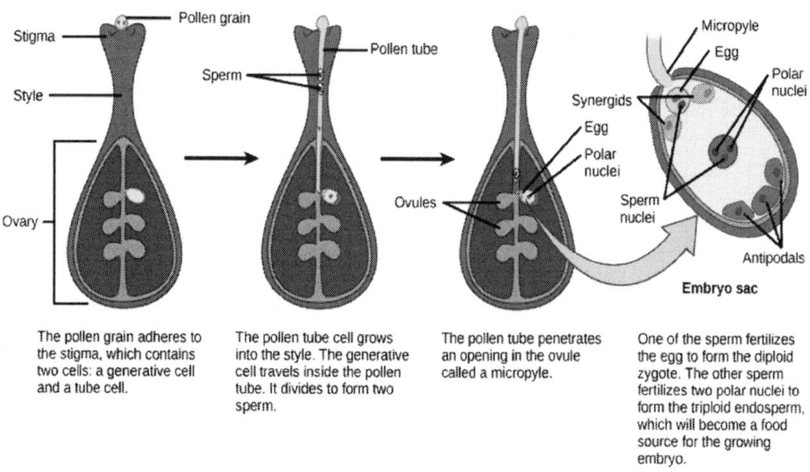

Figure 192: Zygote formation process.

Source: https://biologydictionary.net/zygote/#foobox-1/0/Zygote-process.jpg

Because fungi have slightly varied lifecycles, the process of producing a zygote is slightly different. The vast majority of fungi are haploid. Haploid cells can sometimes unite through a process known as karyogamy, or nuclei fusing (Van der Kooi & Ollerton, 2020). For a brief period of time, this results in a diploid organism. Important genetic diversity and recombination can occur during this time, allowing the fungus to stay resilient in the face of challenges.

BIBLIOGRAPHY

1. Arruda, S. C. C., Silva, A. L. D., Galazzi, R. M., Azevedo, R. A., & Arruda, M. A. Z. (2015). Nanoparticles applied to plant science: a review. *Talanta*, *131*, 693-705.
2. Bourgaud, F., Gravot, A., Milesi, S., & Gontier, E. (2001). Production of plant secondary metabolites: a historical perspective. *Plant science*, *161*(5), 839-851.
3. Chandler, S. F., & Sanchez, C. (2012). Genetic modification; the development of transgenic ornamental plant varieties. *Plant biotechnology journal*, *10*(8), 891-903.
4. Craigie, J. S. (2011). Seaweed extract stimuli in plant science and agriculture. *Journal of applied phycology*, *23*(3), 371-393.
5. Clemens, S., Aarts, M. G., Thomine, S., & Verbruggen, N. (2013). Plant science: the key to preventing slow cadmium poisoning. *Trends in plant science*, *18*(2), 92-99.
6. Chen, L., & Yamaguchi, S. (2002). Genetic diversity and phylogeny of tea plant (*Camellia sinensis*) and its related species and varieties in the section Thea genus Camellia determined by randomly amplified polymorphic DNA analysis. *The Journal of Horticultural Science and Biotechnology*, *77*(6), 729-732.
7. De Waele, D., & Elsen, A. (2007). Challenges in tropical plant nematology. *Annu. Rev. Phytopathol.*, *45*, 457-485.
8. Elmer, W., & White, J. C. (2018). The future of nanotechnology in plant pathology. *Annual review of phytopathology*, *56*, 111-133.

9. Fahad, S., Bajwa, A. A., Nazir, U., Anjum, S. A., Farooq, A., Zohaib, A., ... & Huang, J. (2017). Crop production under drought and heat stress: plant responses and management options. *Frontiers in plant science*, *8*, 1147.
10. Foxcroft, L. C., Pyšek, P., Richardson, D. M., Genovesi, P., & MacFadyen, S. (2017). Plant invasion science in protected areas: progress and priorities. *Biological Invasions*, *19*(5), 1353-1378.
11. Franklin, J., Davis, F. W., Ikegami, M., Syphard, A. D., Flint, L. E., Flint, A. L., & Hannah, L. (2013). Modeling plant species distributions under future climates: how fine scale do climate projections need to be? *Global change biology*, *19*(2), 473-483.
12. Guo, Z., & Liu, W. (2007). Biomimic from the superhydrophobic plant leaves in nature: Binary structure and unitary structure. *Plant Science*, *172*(6), 1103-1112.
13. Hadwiger, L. A. (2013). Multiple effects of chitosan on plant systems: Solid science or hype. *Plant science*, *208*, 42-49.
14. Hirschberg, J. (2001). Carotenoid biosynthesis in flowering plants. *Current opinion in plant biology*, *4*(3), 210-218.
15. Jeger, M. J., Pautasso, M., Holdenrieder, O., & Shaw, M. W. (2007). Modelling disease spread and control in networks: implications for plant sciences. *New Phytologist*, *174*(2), 279-297.
16. Katova, A. (2007). Species and varieties of perennial grasses for high quality forage in Bulgaria. *Journal of Mountain Agriculture on the Balkans*, *10*(1), 156-161.
17. Klassen, R. D. (2001). Plant-level environmental management orientation: the influence of management views and plant characteristics. *Production and Operations Management*, *10*(3), 257-275.
18. Lambers, H., Chapin, F. S., & Pons, T. L. (2008). *Plant physiological ecology* (Vol. 2). New York: Springer.
19. Luan, Y., Fu, X., Lu, P., Grierson, D., & Xu, C. (2020). Molecular mechanisms determining the differential accumulation of carotenoids in plant species and varieties. *Critical Reviews in Plant Sciences*, *39*(2), 125-139.
20. Lindström, B. E. M., Frankow-Lindberg, B. E., Dahlin, A. S., Wivstad, M., & Watson, C. A. (2013). Micronutrient concentrations in common

and novel forage species and varieties grown on two contrasting soils. *Grass and Forage Science*, *68*(3), 427-436.

21. Mandel, J. R. (2019). A Jurassic leap for flowering plants. *Nature plants*, *5*(5), 455-456.
22. Matyjaszczyk, E., & Śmiechowska, M. (2019). Edible flowers. Benefits and risks pertaining to their consumption. *Trends in Food Science & Technology*, *91*, 670-674.
23. Mercuri, A., Sacchetti, A., De Benedetti, L., Schiva, T., & Alberti, S. (2002). Green fluorescent flowers. *Plant Science*, *162*(4), 647-654.
24. Novello, N., Ferfuia, C., Pasković, I., Fabris, A., Baldini, M., Schat, H., & Pošćić, F. (2020). Independent variation in copper tolerance and copper accumulation among crop species and varieties. *Plant Physiology and Biochemistry*, *156*, 538-551.
25. Nikunj, P., Purvi, D., Niti, P., Anamika, J., & Gautam, H. K. (2014). Agronanotechnology for plant fungal disease management: a review. *International Journal of Current Microbiology and Applied Sciences*, *3*(10), 71-84.
26. Neumann, G., George, T. S., & Plassard, C. (2009). Strategies and methods for studying the rhizosphere—the plant science toolbox. *Plant and Soil*, *321*(1), 431-456.
27. Prajapati, K., & Modi, H. A. (2012). The importance of potassium in plant growth–a review. *Indian Journal of Plant Sciences*, *1*(02-03), 177-186.
28. Pauchard, A., Kueffer, C., Dietz, H., Daehler, C. C., Alexander, J., Edwards, P. J., & Seipel, T. (2009). Ain't no mountain high enough: plant invasions reaching new elevations? *Frontiers in Ecology and the Environment*, *7*(9), 479-486.
29. Poff, N. L., & Zimmerman, J. K. (2010). Ecological responses to altered flow regimes: a literature review to inform the science and management of environmental flows. *Freshwater biology*, *55*(1), 194-205.
30. Reid, A. M., Morin, L., Downey, P. O., French, K., & Virtue, J. G. (2009). Does invasive plant management aid the restoration of natural ecosystems? *Biological Conservation*, *142*(10), 2342-2349.
31. Rusman, Q., Lucas-Barbosa, D., Poelman, E. H., & Dicke, M. (2019). Ecology of plastic flowers. *Trends in plant science*, *24*(8), 725-740.

32. Simon, S., & Petrášek, J. (2011). Why plants need more than one type of auxin. *Plant Science*, *180*(3), 454-460.
33. Smith, M. D., & Knapp, A. K. (2001). Physiological and morphological traits of exotic, invasive exotic, and native plant species in tallgrass prairie. *International Journal of Plant Sciences*, *162*(4), 785-792.
34. Schuurink, R. C., Haring, M. A., & Clark, D. G. (2006). Regulation of volatile benzenoid biosynthesis in petunia flowers. *Trends in plant science*, *11*(1), 20-25.
35. Singh, R. (2015). Medicinal plants: A review. *Journal of Plant Sciences*, *3*(1), 50-55.
36. Stenberg, J. A. (2017). A conceptual framework for integrated pest management. *Trends in plant science*, *22*(9), 759-769.
37. Szpak, P. (2014). Complexities of nitrogen isotope biogeochemistry in plant-soil systems: implications for the study of ancient agricultural and animal management practices. *Frontiers in plant science*, *5*, 288.
38. Taiz, L., Zeiger, E., Møller, I. M., & Murphy, A. (2015). *Plant physiology and development* (No. Ed. 6). Sinauer Associates Incorporated.
39. Twidwell, D., Rogers, W. E., Fuhlendorf, S. D., Wonkka, C. L., Engle, D. M., Weir, J. R., ... & Taylor Jr, C. A. (2013). The rising Great Plains fire campaign: citizens' response to woody plant encroachment. *Frontiers in Ecology and the Environment*, *11*(s1), e64-e71.
40. Van der Kooi, C. J., & Ollerton, J. (2020). The origins of flowering plants and pollinators. *Science*, *368*(6497), 1306-1308.
41. Van Breusegem, F., Vranová, E., Dat, J. F., & Inzé, D. (2001). The role of active oxygen species in plant signal transduction. *Plant science*, *161*(3), 405-414.
42. Van Bruggen, A. H. C., & Finckh, M. R. (2016). Plant diseases and management approaches in organic farming systems. *Annual review of phytopathology*, *54*, 25-54.
43. Van Dijk, A. D. J., Kootstra, G., Kruijer, W., & de Ridder, D. (2021). Machine learning in plant science and plant breeding. *Iscience*, *24*(1), 101890.
44. Waraich, E. A., Ahmad, R., Ashraf, M. Y., Saifullah, & Ahmad, M. (2011). Improving agricultural water use efficiency by nutrient management in crop plants. *Acta Agriculturae Scandinavica, Section B-Soil & Plant Science*, *61*(4), 291-304.

45. Wilkinson, S., Kudoyarova, G. R., Veselov, D. S., Arkhipova, T. N., & Davies, W. J. (2012). Plant hormone interactions: innovative targets for crop breeding and management. *Journal of experimental botany*, *63*(9), 3499-3509.

46. Weitemier, K., Straub, S. C., Cronn, R. C., Fishbein, M., Schmickl, R., McDonnell, A., & Liston, A. (2014). Hyb-Seq: Combining target enrichment and genome skimming for plant phylogenomics. *Applications in plant sciences*, *2*(9), 1400042.

47. Wisler, G. C., & Norris, R. F. (2005). Interactions between weeds and cultivated plants as related to management of plant pathogens. *Weed Science*, *53*(6), 914-917.

48. Yakhin, O. I., Lubyanov, A. A., Yakhin, I. A., & Brown, P. H. (2017). Biostimulants in plant science: a global perspective. *Frontiers in plant science*, *7*, 2049.

49. Yakhin, O. I., Lubyanov, A. A., Yakhin, I. A., & Brown, P. H. (2017). Biostimulants in plant science: a global perspective. *Frontiers in plant science*, *7*, 2049.Clemens, S., Aarts, M. G., Thomine, S., & Verbruggen, N. (2013). Plant science: the key to preventing slow cadmium poisoning. *Trends in plant science*, *18*(2), 92-99.

50. Yamada, T., Takatsu, Y., Manabe, T., Kasumi, M., & Marubashi, W. (2003). Suppressive effect of trehalose on apoptotic cell death leading to petal senescence in ethylene-insensitive flowers of gladiolus. *Plant Science*, *164*(2), 213-221.

51. Zehnder, G., Gurr, G. M., Kühne, S., Wade, M. R., Wratten, S. D., & Wyss, E. (2007). Arthropod pest management in organic crops. *Annu. Rev. Entomol.*, *52*, 57-80.

INDEX